EUROPEAN ASSOCIATION OF GEOGRAPHERS

Book Series

GEOGRAPHY : OUR WORLD!

1

Bibliografische Information der *Deutschen Nationalbibliothek*
Die Deutsche Nationalbibliothek verzeichnet diese Publikation in der Deutschen Nationalbibliografie; detaillierte bibliografische Daten sind im Internet über <http://dnb.ddb.de> abrufbar.

ISBN: 978-3-86387-470-4

EUROGEO
GEOGRAPHY: OUR WORLD! - Bd. 1

Dieses Werk ist urheberrechtlich geschützt.
Alle Rechte, auch die der Übersetzung, des Nachdruckes und der Vervielfältigung des Buches, oder Teilen daraus, vorbehalten. Kein Teil des Werkes darf ohne schriftliche Genehmigung in irgendeiner Form reproduziert oder unter Verwendung elektronischer Systeme verarbeitet, vervielfältigt oder verbreitet werden.

Die Wiedergabe von Gebrauchsnamen, Warenbezeichnungen, usw. in diesem Werk berechtigt auch ohne besondere Kennzeichnung nicht zu der Annahme, dass solche Namen im Sinne der Warenzeichen- und Markenschutz-Gesetzgebung als frei zu betrachten wären und daher von jedermann benutzt werden dürfen.

This document is protected by copyright law.
No part of this document may be reproduced in any form by any means without prior written authorization of the publisher.

Coverabbildung: © Robert Kotsch - Fotolia.com

© **mbv**berlin - Mensch und Buch Verlag 2014
Choriner Str. 85 - 10119 Berlin
verlag@menschundbuch.de – www.menschundbuch.de

EUROPEAN ASSOCIATION OF GEOGRAPHERS

GEOSPATIAL TECHNOLOGIES AND GEOGRAPHICAL ISSUES

Edited by

Kostis C. Koutsopoulos
and
Yorgos N. Photis

PREFACE

For the presidium of EUROGEO, the European Association of Geographers, it is a great pleasure as well as an immense satisfaction to see the first book of the book series titled "Geography: Our World!" materialize. Indeed this is a turning point between a process that started several years ago with the formation of the Association of European Geographers and a new era whereby our Association has started publishing the European Journal of Geography and now the this book series.

There is plenty on the news about what Geography and Geographers are doing, but there are precious few references for explaining Geography's role in approaching everyday issues to audiences who mostly need to know, namely decision makers. To fill this gap in the literature and examine specifically society's needs, conditions and changes from a spatial perspective, EUROGEO launches this book series.

Each book of the series will present a collection of papers on topics that reflect the significance of Geography as a discipline. That is, the overall aim is to deliver thought-provoking contributions that explore the complex interactions among geography, technology, politics and the human conditions. In addition these volumes will be brief, clear and to the point, while at the same time tackling urgent topics to geographers and politicians alike.

TABLE OF CONTENTS

	Page
Editorial	1

DIGITAL-EARTH: A EUROPEAN NETWORK IN GEOGRAPHY 4
Karl DONERT

„GLO*KAL* CHANGE": GEOGRAPHY MEETS REMOTE SENSING IN THE CONTEXT OF THE EDUCATION FOR SUSTAINABLE DEVELOPMENT 16
Markus JAHN, Michelle HASPEL and Alexander SIEGMUND

REDEFINITION OF THE GREEK ELECTORAL DISTRICTS THROUGH THE APPLICATION OF A REGION-BUILDING ALGORITHM 27
Yorgos N. PHOTIS

MEASURING EQUITY AND SOCIAL SUSTAINABILITY THROUGH ACCESSIBILITY TO PUBLIC SERVICES BY PUBLIC TRANSPORT. THE CASE OF THE METROPOLITAN AREA OF VALENCIA (SPAIN) 37
María-Dolores PITARCH GARRIDO

"DTh 1.0": TOWARDS AN ARTIFICIAL INTELLIGENCE DECISION SUPPORT SYSTEM FOR GEOGRAPHICAL ANALYSIS OF HEALTH DATA 56
Dimitris KAVROUDAKIS and Phaedon C. KYRIAKIDIS

MAPPING GEOGRAPHICAL INEQUALITIES OF INFORMATION ACCESSIBILITY AND USAGE: THE CASE OF HUNGARY 66
Ákos JAKOBI

DYNAMIC OPPORTUNITY-BASED MULTIPURPOSE ACCESSIBILITY INDICATORS IN CALIFORNIA 78
Pamela DALAL, Yali CHEN and Konstadinos G. GOULIAS

REDEVELOPING THE GREYFIELDS WITH ENVISION: USING PARTICIPATORY SUPPORT SYSTEMS TO REDUCE URBAN SPRAWL IN AUSTRALIA 89
Stephen GLACKIN

EDITORIAL

By its nature Geography connects the physical, human and technological sciences enhancing teaching, research, and of interest to decision makers, problem solving. As a result, Geography provides answers of how aspects of these sciences are interconnected and are forming spatial patterns and processes that impact on global issues and thus effecting present and future generations. Moreover, Geography by dealing with places, people and cultures, it explores international issues ranging from physical, urban and rural environments and their evolution, to climate, pollution, development and political-economy.

In addition, the development of Geospatial Technologies (GT) has created a profound revolution in science and technology as well as in information access and spatial planning. That is, the acquisition and use of geospatial information, combined with developments in computing and communications has made GT an invaluable tool to scientists, planners and decision makers. In other words, GT play an increasingly important role in addressing the social, economic, cultural, scientific, and technological challenges affecting the way we understand the world we live on. As a result, GT allows decision makers to focus their attention on many of the important challenges society is facing nowadays, such as economic efficiency, resource depletion, sustainable energy, natural hazards, food and water supplies, environmental degradation, population migration as well as the many problems in our cities. Therefore, the recent developments of Geospatial Technologies can be used to bridge the gap between citizens, decision makers and real-world problems by providing the necessary tools and processes. That is, GT can be used and are extremely important because:

- They can contribute to a better understanding of universal problems and their interaction
- They can provide decision makers with the knowledge and skills needed in order to understand and cope with or even resolve these problems
- They can stimulate or revive interests leading into action.

Based on these, decision makers, and politicians in particular, need to have an understanding of the ways in which geospatial technologies can enhance their efforts in dealing with societal problems. That is, Geospatial Technologies, which represent an expression of IT approaches, are very appropriate endeavours to harness the benefits of ICT for society, by utilizing the spatial point of view of events, phenomena, places and of course of the physical and human environment which they represent.

In response to that necessity the aim of this book, which is a selection of papers from the European Journal of Geography (EJG), is to highlight the contribution of geographic research, in the form of Geospatial Technologies applications, in dealing with critical everyday issues such as health, education, accessibility etc. Practically every issue of the EJG shows that modern Geography endowed with a wide variety of geospatial technologies is able to efficiently address those issues and provide decision makers with effective means to successfully approach them.

The publication of the European Journal of Geography is based on the European Association of Geographers' goal to make European Geography a worldwide reference and standard. As a result, the papers published in the EJG are focused in promoting the significance of geography as a discipline in resolving global issues or applying geography, complementing, of course, the fundamental goals of improving the quality of research, learning and teaching of Geography. In other words with the EJG the European Association of Geographers provides a forum for geographers worldwide to communicate on all aspects of research and applications of geography with a European dimension, but not exclusive.

As a result, the book you have in your hands provides a glimpse of the important role Geospatial Technologies can play in helping decision makers and politicians, among others, in resolving everyday problems. More specifically, the book contains eight papers from articles already published on the European Journal of Geography presenting applications of Geospatial Technologies which is hoped to result in the enhancement of the spatial understanding and geographical reasoning of the book readers.

All papers are focused on the application of tools that Geospatial Technologies offer for facing real, every day problems. In the majority of the papers their applications are in various countries of Europe, but two of them represent efforts in other continents to show that the application of GT tools, the problems requiring such tools, the role of Geography and the contributions to the European Journal of Geography transect national boundaries, cultures, political systems and scientific backgrounds.

The papers included in the book and their contribution in providing solutions to existing problems follows:

The paper by **K. DONERT**, the leading paper of the book, shows that Geospatial Technologies are now employed in analysing, understanding and planning many aspects of the world we live in. Moreover, the paper clearly indicates that the Digital Earth approach integrates a range of geospatial technologies used in dealing with many geographical issues and in particular in Geographic education. It is therefore surprising that European school education has so far, by and large, ignored geospatial developments. The challenge then, described in this contribution, is how to scale these developments up through networking initiatives. The result and the focus of the paper has been the establishment of the digital-earth.eu project which connects organisations involved in geospatial education in order to share practice, provide advice and guidance on the use of geographic information to others and be a place for new initiatives.

M. JAHN, **M. HASPEL** and **A. SIEGMUND** contribution presents the web-based learning platform „GLOKAL Change" which provides several interactive learning modules, addressed to young students aged 10 to 16 years in order
to help them learn to evaluate economic, ecological and social impacts of recent environmental changes occurring in different geographic areas. More specifically, by providing information from remote sensing data and other media, users compare, visually analyze and interpret satellite imagery to obtain spatial information on the development of the three dimensions of sustainability.

The paper by **Y. PHOTIS** using the constrained-based spatial clustering algorithm SPiRAL (Spatial Integration and Redistricting ALgorithm), formulates a methodological approach for the definition of homogenous spatial clusters, taking into account both geographical and descriptive characteristics. The approach was used to solve a realistic electoral redistricting problem, that of redefining the electoral districts of the Prefecture of Lakonia in Greece.

M. D. PITARCH GARRIDO paper proposes an approach to spatial equity in complex spaces such as metropolitan areas, based on the population's access to essential public services (education, health care and social services). Using Geographic Information System (GIS) tools to handle both spatial and statistical data the study, which covers the Metropolitan Area of Valencia in Spain, aims at providing a general overview of the situation, at pointing out problem zones, as well as suggesting answers for them.

The contribution of the paper by **D. KAVROUDAKIS** and **P. C. KYRIAKIDIS** is the development of an autonomous Decision Support System for a real-time analysis of health data with the use of decision trees. The system using individual patients' datasets, based on their symptoms and other relevant information, prepares reports about the importance of the characteristics that determine the number of patients of a specific disease. That is, the paper describes the design of a tree-based system and uses a virtual database to illustrate the classification of patients in a hypothetical intra-hospital case study.

Á. JAKOBI paper sets up a theoretical framework and applied it, using multivariable analysis of information accessibility differences, to Hungarian micro regions in order to determine the physical infrastructural constraints on the information society and economy. The results show that access to the new information channels, a necessity to all, should not be treated as a secondary problem. That is, when dealing with the newest geographical inequalities of the information age, the inequalities in the quality of information usage is more important than the usage volume differences.

The paper by **P. DALAL**, **Y. CHEN** and **K. G. GOULIAS** illustrates an original approach in creating realistic space-sensitive and time-sensitive spatial accessibility indicators based on availability of opportunities. These indicators were created to support the development of the Southern California Association of Governments activity-based travel demand forecasting model that aims at a second-by-second and parcel-by-parcel modeling and simulation.

S. GLAKIN's paper deals with ENVISION a GIS-based, Participatory Support System, for engaging with the diverse array of stakeholders involved in urban redevelopment. The system was designed to bring wide-ranging land, demographic and market data together in order to highlight the redevelopment options and identify potential redevelopment precincts, across metropolitan centers. As a result, the aim of the system was to initiate debate between those involved on how best to manage urban growth. The application of the ENVISION program in Australia has shown that GT tools can have considerable affect in many fronts, such as the mutual education of stakeholders in extracting the pertinent issues and potential barriers to redevelopment and in encouraging groups of experts to produce novel solutions to difficult problems that they themselves could not, without the collaboration that the tool demands, resolve on their own.

The Editors

DIGITAL-EARTH: A EUROPEAN NETWORK IN GEOGRAPHY

Karl DONERT
President, European Association of Geographers, http://www.eurogeography.eu
Director: European Centre of Excellence: digital-earth.eu, 30 Schillerstrasse, A5020 Salzburg, Austria, http://www.digital-earth.eu

Abstract

The Digital Earth approach integrates a range of geospatial technologies used in dealing with many geographical issues. Many hundreds of thousands of people are now employed to use these technologies in analysing, understanding and planning many aspects of the world we live in. It is therefore surprising that European school education has so far, by and large, ignored geospatial developments. Small pockets of intense geospatial education activity had been identified. The challenge described in this paper is how to scale these developments up through networking initiatives. The digital-earth.eu project was established to connect organisations involved in geospatial education, to share practice, provide advice and guidance on the use of geographic information to others and be a place for new initiatives. The project raised awareness of the role digital earth education should play, informing politicians and decision makers of the significance of digital earth tools and technologies. A teacher support network infrastructure was created incorporating geo-services for teachers such as ArcGIS Online for Organisations and the European Environment Agency EyeonEarth platform. The outcomes indicate that Digital Earth education and training developments are urgently needed as part of the European Qualifications Framework. European policy makers have to be made much more aware of geospatial concepts and then actively encouraged by stakeholders to respond to them in policy terms through developing a "Digital Earth education for all".

Keywords: *ArcGIS Online, digital-earth.eu, Digital Earth, geoinformation tools, geo-ICT, lobbying, geospatial, network, geographical education*

1. INTRODUCTION

In 1992, former US VP Al Gore presented a farsighted Digital Earth concept, whereby detailed geospatial information could be accessed from any place, at anytime, by anyone (Gore, 1992). The subsequent scientific and technological movement has made this vision a reality today as geospatial technologies used to research, develop and plan many of the complex issues facing society today. Based on a US Dept of Labor study, Gewin (2004), writing in the scientific publication Nature, proposed that geo-technology (with related spatial thinking skills) would become one of three most significant technological advances for economic development in the next decade. Since then, in the United States there has been a strong lobby for geospatial education, resulting in Congress acknowledging the significance of the National Academies Press publication „Learning to Think Spatially" (National Research Council, 2006). This has transformed the US research and education technology agenda and as a result the National Science Foundation (National Science Foundation, 2011) recently awarded significant grants to geospatial education research. In sum it is now widely accepted that Geospatial technology is imperative in dealing with geographic issues

Developments in Europe, however, have not matched these initiatives as only a few national and European activities, largely based on pilot projects, have been supported and local schemes tended to be small-scale and without such political backing (Jekel et al., 2010). In response to the need to use geospatial technology in dealing with geographic issues and especially in geographic education a European network for digital earth education in geography has been created and is the focus of this paper.

More specifically, the Digital Earth vision connected groups of scientists interested in cooperative studies of the planet and its resources, and directed actions for solutions towards sustainable development. Since then Digital Earth technologies created a profound revolution in science and technology, information access and spatial planning. The acquisition and use of geospatial information, combined with developments in computing and communications has made information about the earth available to billions of people. The Digital Earth concept has become a reality (Gore, 1998) and the results should play an increasingly important role in addressing the social, economic, cultural, scientific, and technological challenges affecting the way we understand the earth. Global agencies like the United Nations have established expert groups to coordinate, dialogue, advise and inform on geospatial matters (UN News, 2011). Specifically, Digital Earth allows us to focus the attention on many of the important challenges faced by Europe today, such as economic efficiency, resource depletion, sustainable energy, natural hazard mitigation, food and water supplies, environmental degradation, population migration and smart cities.

Guo et al. (2010) describe the movement from awareness of Digital Earth to putting its tools and technologies into practice. They comment on how Digital Earth developments led to the integration of different geospatial technologies. For example remote sensing, sensors, geographic information systems, GPS, simulation and virtual reality through web-based developments used to support important infrastructure developments. They also discuss how Digital Earth has become one of the leading technologies supporting science and technology research, particularly used to address global challenges. Recent developments in geographic media (or geo-media) are being used to bridge the gap between citizens, Digital Earth technologies and real-world problems by socially connecting them through geographic location services.

In education, geo-media has the capacity to create powerful learning opportunities that could empower students and through the Cloud, individualise learning. Despite this potential, European education, for instance in science, history, economics, geography, social studies, media and ICT, has so far, by and large, ignored these developments. This is despite the fact that geo-technology has become a significant employer and geoinformation and geo-media have become almost ubiquitous commodities accessible from mobile, tablet and laptop. In schools, the use of geo-media can help students construct concepts and promote a meaningful understanding of our world through problem solving, experimentation, project work and the communication of findings to others (Gryl, Jekel and Donert, 2010). The visual elements offered by geo-media are essential for enquiry, exploration and communication. However, network research confirmed there were only small pockets of intense activity (Lindner-Fally and Donert, 2011) and that geo-media education and the use of geospatial technologies in European schools and teacher training has generally lagged behind (Donert, 2010). This paper reports on an initiative to address this by establishing, developing and promoting the concept of Digital Earth education for schools and in teacher education and training across Europe.

2. CREATING A DIGITAL-EARTH.EU EUROPEAN NETWORK

2.1. Developing a Digital Earth European Centre

Research has confirmed that in Europe there had been little or no attention paid to the significance of emerging geospatial technologies in schools (Milson et al., 2012; Gaudet and Annulis, 2003). A few pilot projects have created teaching resources in several languages and training courses have successfully been delivered to relatively small numbers of educators across the continent. Large-scale, ministerial-initiated implementation was generally lacking, indicating that European education has generally been unable to keep pace with technological and societal changes taking place.

In response to geo-spatial developments 2009 an Austrian Centre for geo-media education (digital:earth:at) was created in Salzburg (Lindner-Fally, 2009), linking a number of organisations who were working with schools and teachers. The goal was to share resources, tools and innovative ideas to increase the use of geo-media with Austrian pupils and teachers. Its successful implementation resulted in the development of a proposal with the European Association of Geographers (EUROGEO) for a European networking initiative, called digital-earth.eu, connecting stakeholders across the continent. A network consisting of 47 partners from 18 countries was formed and funding obtained for them to work together for three years (2010-2013) under the Lifelong Learning Comenius Programme. The digital-earth.eu Comenius network aims to raise awareness of the many innovative ‚geospatial' developments taking place and reflect on their implication and potential impact in school education systems.

An initial focus of the digital-earth.eu project was the founding of a European Centre in November 2011, based at the Austrian Centre of Excellence with the aim to build a Community of Practice that could support teachers in different parts of Europe, and connect people working in national and regional contexts (Jekel et al., 2011). The purpose was to generate a European infrastructure that would allow those involved to: i) share ideas and information, ii) communicate future visions and iii) develop an informed Community of Practice (CoP). The CoP would be based on the development of a network of accredited expert centres for geo-media across Europe.

Following an open Call for experts, the evaluation of proposed Centres of Excellence was undertaken through a peer review process and accredited by the European Centre and EUROGEO. These expert centres form multipliers by working with many teachers and trainers in their own situations. They are able to offer advice and guidance to stakeholders such as Ministries of Education. This process offers increased visibility to organisations that are doing outstanding work. It encourages and supports innovation in learning and teaching approaches and rewards quality. At the time of writing this article, fourteen Centres in thirteen European countries have been established and several others are going through the review process.

2.2. Special Interest Groups (SIGs)

The issues dealt with by the digital-earth.eu project are very broad, including teacher training standards, professional development and geomedia competences. They have considered issues of data availability following the results of the EU INSPIRE initiative and the tools available for educators to use. The network was organised into four thematic special interest groups (SIGs) concerned with:

SIG 1: Data, Tools and Technologies
SIG 2: Learning and teaching environments
SIG 3: Teacher Education and Training
SIG 4: Curriculum developments

Each of these groups have reviewed the state of the art in their area and contributed to an online catalogue of materials, courses, publications, links and best practice scenarios. They have also produced a series of research papers, publications and guidance materials. These keep teachers up-to-date with developments, resources and advice.

A needs analysis of network partners showed that while technical advances have extended the Digital Earth vision in scientific terms (Foresman, 2008; Gore, 1998), in education their uses are still mostly restricted to a few users within schools and teacher training. There has been an explosion in the number of geospatial Web 2.0 tools available for teachers to use with their students, yet digital earth technologies were not widely described in national curricula. Most European Ministries of Education and even the European Commissioners for Education and the Digital Agenda were largely unaware of their existence (Lindner-Fally and Donert, 2011). As a result, it is almost impossible for most teachers to keep pace with the plethora of technologies at their disposal. The Data, Tools and Technologies SIG identified many of these resources and promoted their availability in school and teacher training contexts. These included social media, media content like RSS feeds, blogs and video clips, open apps freely available to download for mobile devices (Al-Khudhairy, 2010), mashup interfaces that allow interactive on-the-fly mapping, sophisticated visualisations and geo-collaborative activities developed via distributed Cloud-based, Web GIS (Alexander, 2006).

SIG 1 explored some educational perspectives of the outcomes of the European INSPIRE initiative and examined the possible impacts for teaching in schools and in teacher education. Members of the group considered data availability, standards and interoperability and addressed property rights from a school perspective, producing advice to inform teachers and teacher educators. It resulted in a series of recommendations for action (Donert, 2013a). A report was also produced which explored issues associated with the European INSPIRE Directive (Kotsev, Filipov and Donert, 2013), as well as copyright, Intellectual Property, standardisation and quality issues concerning data and information in different European countries relating to schools and teachers (Donert, 2013b). Volunteered geographic information (Goodchild, 2007) and crowdsourcing (Howe, 2008) were examined as interesting alternatives to traditional information sources from mapping agencies and companies. An online searchable catalogue of resources has been created (http://www.digital-earth.eu/services/database-search.html) which provides an infrastructure through which links to resources, data, information and teaching materials.

Digital earth technologies can be used in education as a medium to encourage enquiry, enhance communication, construct personalized teaching materials, and assist students' self-expression (Beak et al., 2008). The second working group (SIG 2: Learning Environments) looked at learning and teaching issues connected with the use of geo-media in schools. There are many different aspects that can play a determining role in successful learning. Their focus was on student-centred learning approaches, using geo-media in transmissive, dialogic, constructivist and co-constructive ways (Mishra and Koehler, 2006), so that teachers are able to encourage guided enquiry in their classrooms (Crough et al., 2012). The role of digital storytelling opportunities was considered highly significant, encouraged by Web 2.0 tools and communications technologies (Levine, 2010).

Digital Earth in education offers opportunities for meaningful, deep learning experiences in and beyond schools. It contributes to teaching and learning by supporting exploration and experimentation; it improves motivation and learner engagement; and offers the learners more responsibility and control through individual and group communication (Oster–Levinz and Klieger, 2012). The research undertaken confirmed that European teacher education must focus on spatial thinking, so that learners will understand spatial patterns, linkages, and relationships.

SIG 2 has reported on key competences in the use of geo-media, examining the concept of geo-media literacy. It made recommendations for the inclusion of spatial competences, like spatial citizenship (Gryl et al., 2010) as key competences for lifelong learning. The group then undertook a review of learning and teaching approaches and provided practical guidance for teachers and teacher educators. A book publication (in press) will introduce different learning and teaching approaches to teaching with geo-media and geoinformation by examining comparative methods and including exemplars to highlight best practice. This publication will be connected to a conference dealing with aspects of elearning, geomedia and spatial citizenship in teacher education and schools.

The third special interest group addressed the needs of pre- and in-service teacher education. Teachers are key to an effective use of computers in the education system (Zhao et al., 2001). Kerski (2008) discussed the important role teachers play in using key technologies to prepare students to be tomorrow's decision makers, where they are able to tackle local, regional, and global 21st century issues. He suggests developing positive attitudes towards using technology in education is essential. Research by Teo et al. (2007) confirms this and has shown that a teacher's attitude towards new technologies is a major predictor of its successful use.

The Teacher Education and Training report produced by the group reviewed the state of teacher training and geo-media and made recommendations for benchmarking (Lindner-Fally et al., 2012). It confirmed support must be offered to help teachers develop positive attitudes toward computers (Kadijevich and Haapasalo, 2008). To achieve this, the group created a European Centre for teaching and training in geo-media. A business plan was produced to establish an infrastructure of Centres of Excellence across Europe to support teachers and trainers at grassroots level. The group also looked at quality enhancement issues in training and the formulation of an agreed terminology and a benchmark statement for geo-media. Research was also undertaken to report on teacher accreditation across Europe (Lindner-Fally and Zwartjes, 2012) and the opportunities for certification and accreditation in geoinformation. A booklet for teacher training has been produced to offer a checklist and guidance on incorporating geomedia/geoinformation for those training teachers. It deals with in-service training and continuing professional development of teachers.

Educational technology plays an important role in moving from teacher-centred learning activities to student-centred learning activities. Therefore, having trained teachers who are competent in using and managing educational technology is essential (Smakola, 2008). SIG 3 confirmed the main challenge remains to convince education management stakeholders across Europe that the adoption of digital earth tools in their classrooms and training sessions will both enhance the way they work as well as improve their effectiveness as teachers.

The final special interest group (SIG 4: Curriculum Development) has been examining curriculum opportunities for using geo-media and geoinformation in schools Donert et al., 2012). Most teachers have a strong sense of subject identity and are influenced by disciplinary concerns, but as Kerski (2008) suggests, today's main challenges lie with the general structure of our educational systems. Geo-media applications tend to provide cross-curricular opportunities challenging traditional curriculum development. SIG 4 developed a series of case studies of best practice, gathered through the Centres of Excellence and from earlier projects and initiatives to illustrate how to open access to the use of geomedia to pupils and students (Donert and Parkinson, 2013). This publication provided examples in main curriculum areas, including Mathematics, Languages, Science, History, Economics and Geography. It illustrated some techniques used to engage pupils and some of the outcomes from the classroom. The group also produced resources that target curriculum creators and programme developers to advise and guide those involved in developing curricula, creating courses and lessons using geo-media.

3. DIGITAL-EARTH.EU ACHIEVEMENTS

3.1. Partnership, building a Community of Practice

The European Centre has developed the digital-earth.eu network to reach out to other important target audiences across Europe trans-nationally. In 2010 the original partnership consisted of 47 organisations from 18 countries. By early 2013, it had expanded to 89 partners from 22 countries. The digital-earth.eu consortium constitutes different types of organisations and institutions operating in diverse domains and in different ways. The partnership has educational and research organisations (universities), small, medium enterprises, subject thematic associations, teacher training institutions, NGOs and a national Board of Education. These varied consortium organisations are closely connected to different target groups, enabling direct communication with them according to their own contacts and networks. Partners have reached out to particular stakeholders using the most suitable communications channels. Consideration of the specific character of each partner will be an aspect of dissemination planning from an early stage by creating an optimum and effective dissemination planning strategy.

3.2. Dealing with innovation and change

Digital-earth.eu is working in a rapidly changing social and educational environment. The management of change in education will become very significant if we are to embrace Digital Earth environments that encourage personalized learning. It is clear the adoption, adaptation and integration of geo-media in education cannot currently keep pace with the rapid growth of geo-technologies. Projects like digital-earth.eu are essential for the future of the industry if education is to match the rapidly increasing demands for a geospatial workforce. In future school-to-career developments will be needed if geospatial industry development is to be continued and the increasing demand for geo-media professionals can be met.

During the past 30 months, more than 2,900 new geospatial education developments have been identified from the work of the Comenius network and communicated through Twitter (https://twitter.com/digitaleartheu) and via the LinkedIn group (http://www.linkedin.com/groups/digitaleartheu-3761133). These included new innovative tools, data sources, curriculum materials, learning and teaching developments, case studies and publications. It is therefore challenging to provide any 'comprehensive', up-to-date system of in-service support and initial teacher education for activating the potential of geospatial technologies in education.

The network established an infrastructure of Centres of Excellence who offer face-to-face and online modes of delivery to teachers, trainers and educational stakeholders. In response to this, in the future less traditional and even more flexible approaches are likely to be required and accepted, enabled by the creation of generic examples and scenarios that can be applied across countries and contexts. The provision of support, materials and conduct of courses in local languages, reflecting local/national curricula significantly lowers the threshold for the short-term target groups. Working nationally allowed curriculum-relevant contexts to be developed. However, online courses and those undertaken through the Commission Comenius/Grundtvig training database encourages teachers and trainers to collaborate, in an interdisciplinary way and across the borders.

3.3. Open data, open information

Increasingly data, including information from remotely sensed images, are being made freely available and open to citizens (Kroes, 2012) and in new forms. Under INSPIRE and the Digital Agenda for Europe huge volumes of scientific research data is also being made publicly available and the volume of this will be even greater in the future (European Commission, 2012). The geospatial sciences have been leading the way in these endeavours. Many of the new geospatial tools that encourage citizens to act as scientists are being implemented at national, European and global scales (see Earthwatch: http://www.earthwatch.org/ and EyeonEarth: http://www.eyeonearth.org) and relate location-based information to satellite imagery and aerial photography. The digital-earth.eu network has been operating at the fulcrum of this brave new ‚open science' domain, the world of information repositories, curated data and freedom of information. The network focuses on the important need to develop advice and guidance so education is able to cope with the demands of addressing the critical and responsible use of the overload of information being made available in the social networking boom in a critical, analytical and responsible way.

3.4. European added value

Geospatial technologies use location to help make sense of scientific information. Their use, in critical, reflective ways requires the development of spatial thinking skills (Lambert, 2007; National Research Council, 2006). This is increasingly critical for all aspects of life in Europe. These skills are an essential component of lifelong learning. Integrating Digital Earth approaches into school courses supports the formation of a democratic Europe of active, participative and responsible citizens. The digital-earth.eu network thus targets the needs of young people through European teachers and school education systems. European added value lies in the transferability of the materials and resources involved. This is achieved by linking to spatial concepts, geo-social communication and collaboration. The use of geospatial tools in education leads to problem-based rather than place-based approaches (Jahn et al., 2011), which allows materials and resources to be transferred to specific local backgrounds that draw on the individual / collective experience of learners and educators. The work of the digital-earth.eu network has also connected to different school subjects, offering i) learning in interdisciplinary contexts and ii) addressing real world challenges like sustainable development (de Miguel Gonzalez, 2012). Digital Earth education also helps teachers (and students) to understand the diversity that is Europe and the range of complex issues faced by Europeans. The use of data from the European Environment Agency platform 'EyeonEarth' (http://www.eyeonearth.org) adds a European context and online mapping platforms like ArcGIS Online allow visualisation and some analysis of informaiton.

Digital-earth.eu draws on the progress made under the European INSPIRE directive, enhanced by the Digital Agenda for Europe in making public and increasingly private data available to European citizens. Increasingly European educators will be able to contextualise their activities by making use of this publicly available content to customize the learning environments they create to their specific place, school, college, youth organisation. This individualisation of learning, enabled by Cloud-based developments, requires a different paradigm and training perspective, one that is being promoted by Digital Earth education.

3.5. Spreading excellence, exploiting results, disseminating knowledge

The Digital Earth concept provides a very attractive, positive, relatively well-known scientific brand. The digital-earth.eu network has been able to exploit it to widely promote the concept. This familiarity has created a significant marketing advantage, leading to increased awareness and drawing greater attention to the work of the project. The general aims of dissemination activities within digital-earth.eu project have been to:
- widely disseminate the existence of the digital-earth.eu and its special focus on geospatial technologies and Digital Earth education
- propagate knowledge about the work of the project, its specific character, objectives and planned actions and activities to direct target audiences
- develop and maintain a user-friendly project web site to keep the general public and other interested stakeholders informed about the digital-earth.edu project and its results.
- participate in thematically related international meetings, events and conferences and to organise national workshops to inform the educational community and the project's direct target audience about the development of digital-earth.eu and
- disseminate project results and outcomes by other relevant and suitable means.

The digital-earth.eu network is dealing with a highly innovative, rapidly changing subject area in education. Widespread network dissemination has sought to reach as many relevant organisations as possible, including teacher associations, Ministries, academies and other relevant institutions. The goal has been to raise the profile of learning with digital geo-media, encouraging innovative practices and rewarding organisations and individuals displaying 'excellence'. For example, the network has been working collaboratively with European Schoolnet and connecting with policy makers and decision takers, including European Ministries of Education. Since early 2012, digital-earth.eu has been a featured external project on their Scientix Web portal (http://tinyurl.com/bju7mn3). Further discussion over the possibilities to showcase the importance of geospatial technologies in European STEM education are being established, for example by connecting the geospatial industry with the Future Classroom Lab (http://fcl.eun.org/) hosted in Brussels at Schoolnet.

An important purpose of the digital-earth.eu network has been to influence policy makers who had already begun to connect European social and environmental developments to citizens, but not made the link with location-based technologies. Lobbying activities undertaken predominantly by EUROGEO operating as a partner through digital-earth.eu have led to significant political engagement with the EC 'Digital Agenda for Europe' and 'New Skills New Jobs' initiatives. Dissemination activities promoted the incorporation of 'education for digital earth' into regional, national and European educational agenda.

3.6. Centres of Excellence (CoEs)

The vision of digital-earth.eu has been to create an infrastructure of Centres of Excellence in order to offer leadership in the field; provide information and influence to decision makers; develop and deliver services to the teachers, educators and other stakeholders through strategic actions and joint or collaborative activities. In turn, individual CoEs will establish a network in their catchment areas, build a significant online presence and develop lobbying potential for the inclusion of Digital Earth concepts, content and tools in formal and informal education.

The digital-earth.eu Centres of Excellence (http://www.digital-earth.edu.net) have been validated not only based on their expertise and work with teachers and teacher trainers, but also on their potential sustainability in their own local context and environment. In some cases they are housed in public organisations like universities, teacher training institutions and Ministries of Education and Training, but other are in the private or voluntary sector. The network of Centres of Excellence decentralises dissemination and adds value at local, regional and national scales, helping trainers, teachers and educators establish and promote the right sort of learning and teaching culture in their own institutions. In the Comenius network this has provided very positive perceptions at local and national levels and allowed the CoEs to build their own lobbying power through their status.

4. FUTURE PLANS

The main objective of digital-earth.eu has been to address the use of geospatial technologies from a European perspective and bring geospatial researchers, educators and organisations together in order to advise and inform European stakeholders of the significant geospatial developments in education that have taken place in recent years. The arrival of the Cloud has transformed the potential of geospatial technologies into reality. They now offer valuable, freely available tools for increasing inquiry-based spatial education. Introducing Digital Earth applications into teacher education and training will have a significant impact. Digital Earth education, integrated into school curricula, is part of the key to the process (Hauselt and Helzer, 2012), as well as in continuing teacher education courses necessary to help education meet the challenges of Cloud-based learning and teaching.

The European Association of Geographers (EUROGEO) has taken a leading role in developing a three-phase action plan to develop strategic influence through the Centres of Excellence is envisaged from a project-centric perspective through an 'enterprise' phase. So far, the development of the CoE concept has been almost entirely project-based. However it has already attracted information, advice and support from business and industry. As the CoEs begin to produce outcomes and deliverables and share them across the network, they will facilitate the implementation of an effective portfolio of products and services (phase 2). Association development can trigger applications for operating funds and exploration of sponsorship. CoEs will start to evolve into more enterprise-based organisations serving stakeholders and the market as strategic assets providing sustainable leadership, management and services (phase 3). The emphasis throughout is placed on 'expertise' and 'professionalism' as strategic assets.

New curriculum guidelines and a „Centre of Excellence" approach to teacher education and training have been two essential strategies to address the challenges in raising awareness. Education systems in Europe need to adopt more innovative approaches to teacher training and to curriculum frameworks to embrace the rapidly changing scientific, geospatial education landscape. For Digital Earth education to succeed, innovative, scenario-based pedagogies will need to be developed, piloted and employed, together with the use of the right tools to enhance spatial thinking (Jekel et al., 2012). Professional development of teachers and educators is therefore paramount, together with the creation of suitable resources and materials.

The consortium realises that if impact is to be scaled up, lobbying for Digital Earth education must continue nationally and at European level. In addition, teacher education and training needs to switch from predominantly face-to-face to blended and online modes of delivery, using available Cloud technologies to offer anytime, anyplace, anywhere training. The digital-earth.eu partnership believes an accredited 'Centre of Excellence' approach can offer this, provided the training system encourages and allows experts to operate within it. Additionally, teachers have to be treated as professionals who are in control of the support they need to receive. Helping them to become more self-sufficient in determining their professional development needs is also important.

In the future digital-earth.eu must continue to use the latest cloud-based, Web mashup technologies allow geospatial data and remotely sensed information and data to be combined, visualised and explored. New educational scenarios can then be developed and tested. In order to do this, the digital-earth network has negotiated the use of state-of-the-art tools and technologies for its Centres of

Excellence. From March 2013 onwards they have had access to ArcGIS Online for organisations (http://www.esri.com/software/arcgis/arcgisonline) as a platform for sharing information, creating maps and educational products and building apps for schools and teachers to use. Using the EU Environment Agency EyeonEarth platform (http://www.eyeonearth.org), based on ArcGIS Online, offers digital-earth.eu members more opportunities to experiment with Cloud-based learning.

The digital-earth.eu Comenius network has laid a firm platform on which future projects and developments could be based. Several spinoff projects to create teacher education courses and produce support materials have already been initiated and are being developed with EUROGEO and partner organisations. These include iGuess2: GIS in Several Subjects (http://www.iguess.eu), Spatial Citizenship (http://www.spatialcitizenship.org) and I-USE: Statistics in Education (http://www.i-use.eu). A further funding application for a continuation of the digital-earth.eu network was submitted and approved in Autumn 2013. The new project, called School on the Cloud (http://www.schoolonthecloud.eu) will seek to explore the impact of Cloud-based technologies on education. The network has therefore identified methods, approaches and available resources for teaching and learning with geospatial media. It has promoted educational content and collected, validated and widely disseminated it.

5. CONCLUSIONS

Originally education was fundamental to the Digital Earth concept, as Joseph Kerski (2008) commented:

"The Beijing Declaration on the Digital Earth recommended that Digital Earth 'be promoted by scientific, educational and technological communities, industry, governments, as well as regional and international organisations' (Xu and Chen 1999). The declaration emphasised 'understanding the oneness of the Earth and its relevant phenomena.' It called for 'adequate investments and strong support in 'scientific research and development, education and training.'"

However educational perspectives have not received as much attention as other areas. The digital-earth.eu project, as a direct extension of the original Digital Earth initiative, raises awareness of the importance of geo-technologies and stimulates innovative uses of geo-media in schools and education across Europe. The project has attracted considerable interest, an educational researcher community has been built to make advances in curriculum development, learning and teaching approaches, teacher training and the awareness of useful Digital Earth tools and technologies.

The digital-earth.eu network project was founded to raise awareness of geospatial education and inform politicians and Ministries of the significance of digital earth tools and technologies. It has been developed to connect organisations involved in geospatial education, so they can share practice, provide advice and guidance on the use of geographic (geo-)media to others and be a place for innovative future thinking and new initiatives. The rise of geospatial technologies and particularly the growing availability of information and data to the citizen necessitates a reconsideration of the role and structure of 'secondary' education for Europe2020. Few links to everyday, scientific, technical orientated uses of geoinformation and citizen participation have so far been integrated into the education system (Gryl, 2012), for example, spatial (location-based) thinking has so far largely been argued through the use of computer software and along the lines of traditionally organised subject areas. The iGuess Project (http://www.iguess.eu) confirmed that these boundaries severely limit its application in the classroom, largely due to pressures from curriculum-orientated, subject-driven content and limited access to technology.

The Digital Earth education approach implies the need for more creativity in thinking about school systems, structures, topics of study, timetable organisation and the uses of modern technologies, like mobiles and tablets, regularly available as part of our everyday lives. Full integration will require forward thinking from decision makers, who are prepared to take some 'perceived risks' to provide a modern, meaningful and relevant educational experience. Digital-earth.eu also suggests there will be a significant impact on the organization of individual spare time and even voluntary activities (Dunn, 2007). Participatory and community-based approaches such as the posting of information on social networking sites and data collected in the field by citizen scientists are already commonly in use in the voluntary sector. Developing this sort of participatory engagement in the education of young people allows teachers and students to become aware of the power of spatial thinking, geospatial tools and the use of the Web as a communicative and collaborative medium for citizens (young and less young) to engage with, through their involvement in cross-curricular, in-depth, 'capstone-style' projects (Stark and Treuhardt, 2012).

The growing shortages in the geospatial workforce in Europe, the significance of open data and the EU INSPIRE Directive suggests that Digital Earth education and training developments are urgently needed as part of the European Qualifications Framework (European Commission, 2008). European policy makers have to be made much more aware of geospatial concepts (Marsh et al., 2007; Strobl, 2008) and then actively encouraged by stakeholders to respond to them in policy terms9. This calls for support from all stakeholders to help us create meaningful uses of ICT in schools through developing a "Digital Earth education for all".

REFERENCES

Al-Khudhairy, D. H. A. 2010. Geo-spatial information and technologies in support of EU crisis management, International Journal of Digital Earth, 3 (1), 16-30.

Alexander, B. 2006. Web 2.0 – A new wave of innovation for teaching and learning?, Educause Review, 41, http://www.educause.edu/EDUCAUSE+Review/EDUCAUSE Review MagazineVolume41/Web20ANewWaveofInnovationforTe/158042, accessed 10 December 2012.

Beak, Y., Jung, J., and Kim, B. 2008. What makes teachers use technology in the classroom? Exploring the factors affecting facilitation of technology with a Korean sample, Computers & Education, 50, 224-234.

Crough, J., Fogg, L., and Webber, J. 2012. Challenging Opportunities: Integrating ICT in School Science Education. In Issues and Challenges in Science Education Research 263-280, Netherlands, Springer

de Miguel Gonzalez, R. 2012. Geomedia for Education in Sustainable Development in Spain: an experience in the framework of the aims of digital-earth.eu, European Journal of Geography, 3 (3), 44-56

Donert,. K. (ed.) 2010. Using Geoinformation in European Geography education, Rome, International Geographic Union-Home of Geography.

Donert, K. 2013a, Geoinformation, geomedia and data for schools, http://83.164.139.144/fileadmin/deeu_documents/D2.5-geoinformation-v1.pdf, accessed 20 December 2013

Donert, K. 2013b, Geoinformation, information copyright, quality and other issues for schools, http://moodle.eurogeography.eu/mod/resource/view.php?id=877, accessed 20 December 2013

Donert, K. and Parkinson, A. (Eds.), 2013, Geo-media Case Studies in the Curriculum, http://moodle.eurogeography.eu/mod/resource/view.php?id=871, accessed 20 December 2014

Donert, K., Parkinson, A. and Lindner-Fally, M., 2012, Curriculum Opportunities for GeoInformation in Europe, http://83.164.139.144/fileadmin/deeu_documents/D5.1_SIG4-curriculum-report-v3.pdf, accessed 20 December 2013

Dunn, C.E. 2007. Participatory GIS - a people's GIS?, Progress in Human Geography, 31, 616-637.

European Commission, 2008. The European Qualifications Framework for Lifelong Learning (EQF). Office for Official Publications of the European Communities, Luxembourg.

European Commission, 2012. Scientific data: open access to research results will boost Europe's innovation capacity, EC Europa Press Release, http://europa.eu/rapid/press-release_IP-12-790_en.htm, accessed 6 January 2013.

Foresman, T. W., 2008. Evolution and implementation of the Digital Earth vision, technology and society, International Journal of Digital Earth, 1 (1), 4-16.

Gaudet, C., and Annulis, H. 2003. Building the Geospatial Workforce, URISA Journal, 15 (1), 21-30.

Gewin, V. 2004. Mapping opportunities, Nature 427: 376-377.

Goodchild, M.F. 2007. Citizens as sensors: the world of volunteered geography, GeoJournal, 69 (4),

211-221.

Goodchild, M.F. 2009. Neogeography and the nature of geographic expertise, Journal of Location Based Services, 3 (2), 82-96.

Gore, A., 1992. Earth in the Balance: Ecology and the Human Spirit. Boston, Houghton Mifflin.

Gore, A., 1998. The digital earth, http://www.isde5.org/al_gore_speech.htm, accessed 15 October 2012.

Gryl, I. 2012. A Web of Challenges and Opportunities. New research and praxis in geography education in view of current Web technologies, European Journal of Geography, 3 (3), 33-43

Gryl, I., Jekel, T., and Donert, K. 2010. Spatial Citizenship, In Learning with GeoInformation V, eds. Jekel T, Donert K and Koller A, 2-12, Berlin, Wichman Verlag.

Guo, H. D. , Liu, Z. and Zhu, L. W. 2010. ‚Digital Earth: decadal experiences and some thoughts', International Journal of Digital Earth, 3: 1, 31 — 46

Hauselt, P and Helzer; J. 2012. Integration of Geospatial Science in Teacher Education, Journal of Geography, 111 (5), 163-172.

Howe, J. 2008. Crowdsourcing: why the power of the crowd is driving the future of business, New York, McGraw-Hill.

Jekel T, Donert K and Koller A (eds.) 2010. Learning with GeoInformation V, Berlin, Wichman Verlag

Jekel, T., A. Koller, and K. Donert (eds.) 2011. Learning with GeoInformation VI, Heidelberg, Wichmann Verlag.

Jekel, T., Koller, A., and Strobl, J. 2012. Austria: Links Between Research Institutions and Secondary Schools for Geoinformation Research and Practice, In International Perspectives on Teaching and Learning with GIS in Secondary Schools, Milson A, Demirci A, Kerski, J., 27-36, New York, Springer.

Jahn M, Haspel, M. and Siegmund, A. 2011. Glokal change: Geography meets Remote Sensing in the context of the education for sustainable development, European Journal of Geography, 2 (2), 21-34

Kadijevich, DJ., and Haapasalo, L. 2008. Factors that influence student teacher's interest to achieve educational technology standards, Computers & Education, 50, 262-270.

Kerski, J. 2008. The role of GIS in Digital Earth education, International Journal of Digital Earth, 1 (4), 326-346.

Kotsev, A., Filipov, L. and Donert, K., 2013, INSPIRE Directive: impact for teaching, http://83.164.139.144/fileadmin/deeu_documents/D2.2-SIG1-INSPIREreport-v8-final.pdf, accessed 20 December 2013

Kroes, N. 2012. Open data blog of Mrs Kroes, http://blogs.ec.europa.eu/neelie-kroes/tag/open-data, accessed 5 January 2013.

Lambert, D. 2007. Learning to Think Spatially: GIS as a Support System in the K-12 Curriculum, Geographical Education, 20, 79-80.

Levine, A. 2010. 50+ ways to tell a Web 2.0 story, http:// cogdogroo.wikispaces.com/50+Ways, accessed 5 October 2010.

Lindner-Fally M, Digital:earth:at – Centre for Teaching and Learning Geography and Geoinformatics, Proc. HERODOT Conference, Ayvalik, Turkey, http:// www.herodot.net/conferences/Ayvalik/papers/geotech12.pdf, accessed 10 December 2012.

Lindner-Fally, M. and Donert, K. 2011, Needs Analysis Report, digital-earth.eu Network Project, http://83.164.139.144/fileadmin/deeu_documents/D6.1needs-report_f.pdf, accessed 15 December 2013

Lindner-Fally, M., Mira, H., Vieira Silva, D., Carvoeiras, M.L., Lambrinos, N., de Lazaro y Torres M.L., Schmeinck, D., Zwartjes, L. and Donert, K. 2012, Teacher Education and Training and geo-media in Europe: Research Report, http://83.164.139.144/fileadmin/deeu_documents/4_1_report_teachereducation_final.pdf, accessed 20 December 2013

Lindner-Fally, M. and Zwartjes, L. 2012. Learning and teaching with digital earth. Teacher training and education in Europe, In GI_Forum 2012: Geovizualisation, Society and Learning, Jekel, T.,

Car, A. Strobl, J. and Griesebner, G. (Eds.), 272-282, Berlin, Wichmann, Berlin.

Marsh, M., Golledge, R. and Battersby, S.E. 2007. Geospatial Concept Understanding and Recognition in G6 College Students: A Preliminary argument for Minimal GIS, Annals of the Association of American Geographersi, 97 (4), 696-712.

Milson, A. J., Demirci, A. and Kerski, J. J. (Eds.) 2012. International Perspectives on Teaching and Learning with GIS in Secondary Schools, New York, Springer.

Mishra, P. and Koehler, M.J. 2006, Technological Pedagogical Content Knowledge: A Framework for Teacher Knowledge, Teachers College Record, 108 (6), 1017-1054.

National Research Council, 2006. Learning to think spatially. GIS as a Decision-Support System in the K-12 curriculum, Washington DC, National Academies Press.

National Science Foundation, 2011. NSF Geography and Spatial Sciences (GSS) Program, http://www.nsf.gov/funding/pgm_summ.jsp?pims_id=503621, accessed 5 October 2012.

Oster–Levinz, A., and Klieger, A. 2012, How do we know they can do it? Developing TPACK in a pre–service course, International Journal of Learning Technology, 7 (4), 400-418.

Smarkola, C. 2008. Efficacy of a planned behavior model: Beliefs that contribute to computer usage intentions of student teachers and experienced teachers, Computer in Human Behaviour, 24 (3), 1196-1215.

Stark, H-J. and Treuthardt, C. 2012. Switzerland: Introducing Geo-Sensor Technologies and Cartographic Concepts Through the Map Your World Project, In International Perspectives on Teaching and Learning with GIS in Secondary Schools, Milson A, Demirci A, Kerski, J., 255-262, New York, Springer.

Strobl, J. 2008. Digital Earth Brainware, In Geoinformatics paves the Highway to Digital Earth (gi-reports@igf), eds. J. Schiewe, and U. Michel, 134-138, Osnabrück, University of Osnabrück.

Teo, T., Lee, C. B., and Chai, C. S. 2007. Understanding Pre-service teachers' Computer Attitudes: Applying and Extending the Technology Acceptance Model, Journal of Computer Assisted Learning, 24, 128-143.

UN News, 2011. UN Economic and Social Council sets up committee on global geospatial information, UN News Centre, http://www.un.org/apps/news/story.asp?NewsID=39166&Cr=telecom&Cr1=#.UT256Dmtvao, accessed 6 March 2013.

Zhao, Y., Hueyshan, T., and Mishra, P. 2001. Technology: Teaching and Learning: whose computer is it?, Journal of Adolescent and Adult Literacy, 44, 348-355.

„GLO*KAL* CHANGE": GEOGRAPHY MEETS REMOTE SENSING IN THE CONTEXT OF THE EDUCATION FOR SUSTAINABLE DEVELOPMENT

Markus JAHN
University of Education Heidelberg, Department of Geography – rgeo, Czernyring 22/11-12, 69115 Heidelberg, Germany,
http://www.ph-heidelberg.de/en/home.html, jahn@ph-heidelberg.de

Michelle HASPEL
University of Education Heidelberg, Department of Geography – rgeo, Czernyring 22/11-12, 69115 Heidelberg, Germany,
http://www.ph-heidelberg.de/en/home.html, haspel@ph-heidelberg.de

Alexander SIEGMUND
University of Education Heidelberg, Department of Geography – rgeo, Czernyring 22/11-12, 69115 Heidelberg, Germany.
http://www.ph-heidelberg.de/en/home.html, siegmund@ph-heidelberg.de

Abstract

The web-based learning platform „GLOKAL Change" (www.glokalchange.de), which is currently developed at the University of Education Heidelberg, Germany, highlights four topics of environmental changes in terms of sustainable development. In interactive learning modules, adolescents aged 10 to 16 years learn to evaluate economic, ecological and social impacts of recent environmental changes occurring in different geographic areas worldwide and in Germany. Information is provided by remote sensing data and other media. Users compare, visually analyze and interpret satellite imagery to obtain spatial information on the development of the three dimensions of sustainability. Subsequent to the example areas in the modules, learners examine their individual local surroundings at home on a satellite image mosaic of Germany as well as by performing geo-scientific fieldwork on site. „GLOKAL Change" supports an original encounter by providing worksheets and methodology papers on various fieldwork methods, and by the application of a micro-drone for taking their own aerial imagery.

Keywords: *Web-based Learning Platform, Remote Sensing Data, Geography, Education for Sustainable Development, Geoscientific Fieldwork*

1. INTRODUCTION

Human society needs resources for its economic prosperity and social well-being. In the last century, the combination of global population rise and continuous growth of the world economy caused an ever increasing consumption of resources. At the same time, spatial needs for different land-use types such as living space, industry, services, infrastructure and agriculture diminished the area of natural ecosystems on earth. Both processes, provision of resources and reshaping of natural landscapes, have led to profound interventions into the earth's natural state all over the world. Many of these interactions between human society and environment are very complex in structure as they have simultaneous impacts on economic prosperity, ecological equilibrium and social well-being in a geographic region.

In 1987, the Report of the World Commission on Environment and Development *Our Common Future* (available at http://www.un-documents.net/wced-ocf.htm), known as the Brundtland Report, highlighted the environmental and developmental concerns of present human-environment-interactions, whose characteristics are not sustainable regarding the future. As a reaction to the report the United Nations Conference on Environment and Development (UNCED) launched Agenda 21 in 1992. This comprehensive political action program aims at implementing a more sustainable development in the 21st century. To achieve this objective increasing economic effectiveness needs to be combined with more ecologic compatibility and growing social equity as prerequisite to give future generations the same chances to meet their material needs (UN 1993, UNESCO 2011).

One of the most important keys to more sustainable behavior in our society may be a „…reorientation of the education towards more sustainable development…" (Gross & Friese 2000, Bahr 2007). The importance of Education for Sustainable Development (ESD) has been emphasized in chapter 36 of the Agenda 21 (UN 1993). Education for Sustainable Development enables people to „…apply their knowledge on sustainable development and to be aware of the problems of non-sustainable development…", i.e. to identify the mutual dependency of the three dimensions of sustainability as well as to make decisions and act sustainable oneself based on this awareness (cf. Programm Transfer-21 2007, de Haan & Gerhold 2008). In the Lucerne Declaration on Geographic Education for Sustainable Development drafted by Haubrich et al. (2007) the authors stress the necessity „… for the paradigm of sustainable development to be integrated into the teaching of geography at all levels and in all regions of the world.". For Gross & Friese (2000), Hemmer (2006a, 2006b) and Bahr (2007) the subject geography is of importance in the context of ESD due to the analyses of human-environment-interactions and their implications on a geographic area conducted in the subject. Hence, the subject geography is bound to teach for sustainability, i.e. to comprise the concept of ESD in its subject-specific education (DGfG 2010), as almost all topics of the UN Decade of Education for Sustainable Development (UNDESD) 2005-2014 possess a geographic dimension (cf. Haubrich et al. 2007).

In the concept of ESD, education represents a notion of individual competences (BLK 1998, de Haan & Gerhold 2008). Evaluation and media competences for example are both necessary for individuals to comprehend and practice sustainability comprehensively (de Haan & Gerhold 2008, Programm Transfer-21 2007). In Germany, the national educational standards for the school subject geography demand the promotion of both competences: students should be able to obtain information from different media, e.g. satellite imagery, and to evaluate human interventions into the environment concerning their economic, ecologic and social/political compatibility (DGfG 2007). The web-based learning platform (LP) „GLO*KAL* Change" has been developed to focus on fostering both competences: learners are requested to evaluate environmental changes in terms of sustainable development by analyzing given information such as remote sensing (RS) data.

2. THE LEARNING PLATFORM „GLO*KAL* CHANGE"

Fostering students' abilities to comprehend and evaluate the impact of environmental changes on sustainability is the overall aim of the web-based LP „GLO*KAL* Change". Its development has been conducted in the Department of Geography at the University of Education Heidelberg, Germany, as an integral part of the research project „GLO*KAL* Change – Evaluating global environmental changes locally". The LP is still open for use free of charge at www.glokalchange.de and addresses primarily German students from grades 5 to 10 as well as adolescents from extracurricular environmental education aged 10 to 16 years (both described as adolescents later on).

Due to its contribution to the ESD from the viewpoint of geography, the entire project „GLO*KAL* Change" has been marked out as an official project of the UNDESD by the German UNESCO Commission in 2010. The adolescents learn to evaluate environmental changes in terms of sustainability by dealing with each of the three modules of the educational concept of „GLO*KAL*

Change" shown in Figure 1. In interactive learning modules (see section 3) they get to know the impacts of environmental change occurring in different geographic areas on the global (worldwide) and local (in Germany) scale using satellite imagery (see chapter 4). Afterwards, they use a map server containing satellite imagery of Germany to virtually discover their individual home area (see chapter 5). Subsequently, they perform geo-scientific fieldwork on site to explore their local surroundings more comprehensively (see chapter 6). In this context, a micro-drone may be applied assisting the adolescents in the acquisition of further information on their geographic area of interest through taking high resolution aerial imagery (see chapter 6.1).

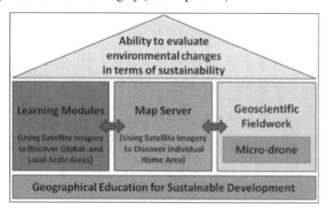

Figure 1. Educational concept of the learning platform „GLO*KAL* Change" to foster adolescents' ability to evaluate environmental changes in terms of sustainability.

3. LEARNING CONTENT AND STRUCTURE OF THE LP „GLO*KAL* Change"

As environmental changes studied in many geographic topics are often complex in structure due to extensive economic, ecologic and social interactions among each other, „GLO*KAL* Change" concentrates on certain geographic topics of human-environment-relationships that are also important to sustainable development. Each topic is presented as an interactive learning module, which has been designed for adolescents in terms of established teaching methods being used in the didactics of geography. Altogether, the LP „GLO*KAL* Change" contains the following four learning modules:

a) The module „Land use" deals with spatial conflicts between economic, ecological and social needs of using space, which have primarily been caused by urban processes, such as the growth of the city of Las Vegas, or suburbanization tendencies in the city of Berlin.
b) The impacts of the cultivation of energy crops and their subsequent transformation to biofuels are discussed in the module „Biofuels from Agriculture", for example the deforestation of the Amazonian tropical rainforest for growing sugarcane to produce ethanol fuel.
c) In „Ecosystem Forest and its Management" the consequences of non-sustainable forest management and forest replacement on ecosystem services have been picked out as a central theme, e.g. the impacts of the deforestation at the airport of Frankfurt am Main, Germany, to build a new runway, or the implications of deforestation in the Congolese rainforest.
d) The effects of mining resources in vast open pits on the economy, environment and society are given attention to in the module „Mining Resources in Open-Cast Mining", for example in the lignite mining area of the Rhineland, Western Germany, or in the Athabasca Oil Sands Area in Alberta, Western Canada.

Each learning module can be accessed from the first page of the LP (Figure 2), they present four examples of non-sustainable environmental change in different geographic areas. In each case, two examples are located on the global scale (worldwide) and at the local scale (in Germany; Figure 3). For every example presented in the modules, the adolescents get to know the recent development to the economic, ecological and social dimension on site as a result of the environmental changes occurring there. They learn to make statements on the dimensions' impact on sustainability. Finally, they are asked to evaluate the whole situation concerning its effects on sustainable development. Users can switch between the learning modules and geographic examples at any time. Each module begins with a short web trailer, which presents the module's topic cinematographically and introduces a set of problems related to the environmental changes occurring within the topic. After the web trailer, general information on the topic is given before users can deal with one of the geographic examples.

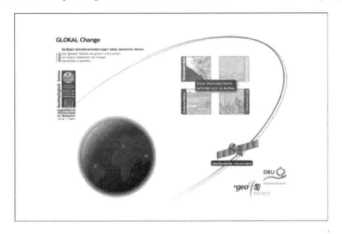

Figure 2. First page of the web-based learning platform „GLO*KAL* Change". Users can enter the four learning modules by clicking on the four images in the middle of the page as well as a map server by clicking on the earth symbol or the small satellite.

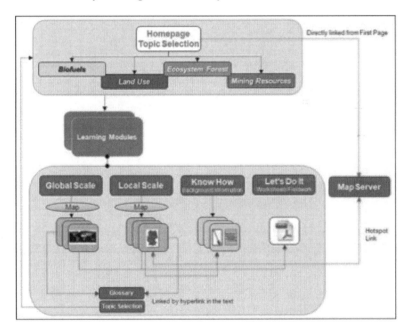

Figure 3. Technical structure of the learning platform „GLO*KAL* Change".

For each example worksheets are available on the LP free of charge. Adolescents can use them as a basis for gathering information on the economic, ecological and social circumstances in the area they deal with. At the same time the worksheets are an opportunity for teaching staff to control the learning progress.

4. UNDERSTANDING SPATIAL IMPACTS BY USING RS DATA IN LEARNING MODULES

Young people can learn about recent economic, ecological and social changes that have occurred in the geographic examples of „GLO*KAL* Change" using a variety of media such as texts, charts, graphics, images, maps and RS data. Satellite imagery is used in the LP for two reasons: first of all, it visualizes spatial relations and changes in the geographic area of interest, which may help learners to understand the impacts on the economy, environment and society. Secondly, the use of satellite imagery induces high motivation and interest by the learners as an international comparative study conducted by Siegmund (2011) revealed. In this study, younger students especially were highly motivated and interested although their specialist knowledge about satellite imagery was generally lower when compared with older participants (Siegmund 2011). In general, factors such as image coloring, image complexity, image ambiguity and general difficulties in image understanding (Gerber & Reuschenbach 2005) often prevent inexperienced users such as younger adolescents from reading and interpreting satellite imagery successfully. For that reason, Beckel & Winter (1989) point to the importance of a gradual image analysis following instruction and/or the provision of additional information, e.g. describing image content. In the LP „GLO*KAL* Change", satellite imagery is embedded into a framework of additional information, which is thought to help users understand and and more deeply interpret the imagery. Moreover, the worksheets described earlier (cf. section 3) contain different exercises in terms of reading and interpreting the imagery correctly. Thus, users are guided step by step towards a comprehensive image interpretation in „GLO*KAL* Change" in order to exhaust the full potential of using RS data in learning situations.

Since the start of the Landsat program in 1972, satellite imagery of the earth's surface is available for more than 30 years, which allows for a time series analysis to get a deep insight into the spatio-temporal development of a geographic area. While Landsat imagery has a low spatial resolution of 30 m × 30 m per pixel, particular RS data such as IKONOS or QuickBird as well as aerial imagery (all less than 1 m × 1 m per pixel) provide detailed information on a geographic area, and permit spatial analysis at a small scale. As a satellite image displays an area covering at least 100 km² or more, depending on the satellite sensor that has been chosen, „…large spatial structures and environmental changes…" (Kollar et al. 2008: 70) can be identified when using these data. Furthermore, satellite imagery displays the earth's structures in its natural appearance in contrast to maps, since no artificial entries or modifications in the imagery have been carried out (Gerber & Reuschenbach 2005, Siegmund & Menz 2005). This is especially true for real color imagery displaying the earth in its natural colors (RGB). False color imagery provides image information on specific subjects of interest such as geologic/geomorphologic features or soil and vegetation properties, and allows for the differentiation of urban and non-urban areas. In general, satellite imagery can help users to make statements on single image objects, connections between these objects as well as on image structures. Thus, reading and interpreting an image correctly can provide a lot of information on the geographic area that has been mapped. The German Educational Standards in Geography for the Intermediate School Certificate point to satellite and aerial imagery as sources for geographic information, which students should be able to acquire (DGfG 2010). Brucker (2006, 178) and Doering & Veletsianos (2007) refer to RS data as suitable tools for the analysis and evaluation of alterations in economic, ecological and social dimensions. This suggests the usefulness of RS data as valuable media in terms of communicating learning content on sustainable development.

According to the explanations made here, the interpretation of RS data is an important method in „GLO*KAL* Change" for gathering spatial information on the kind and extent of changes in the field of the three dimensions of sustainability. The imagery in „GLO*KAL* Change" primarily consists of Landsat TM and ETM+ data. Besides the analysis of single images in real and false color, users can detect spatio-temporal developments or changes within the mapped area by comparing imagery of different temporal origin (Figure 4). In the geographic example „Mining Lignite in the Rhineland, Western Germany", which belongs to the module „Mining Resources in Open-Cast Mining", spatio-temporal changes in this area can be observed in all three dimensions using satellite imagery (Figure

5): for example, the shifting of open-cast mining (economy), the loss of land due to excavation followed by reclamation processes (environment) as well as resettlement activities (social dimension). After the use of the imagery shown in Figure 5, a group of 22 sixth graders (about 12 years old) was asked „Did the satellite imagery help you to understand the topic?" Seventeen answered with „Yes.", five said „It was ok." and none of them answered „No.". In the modules „Biofuels from Agriculture" and „Ecosystem Forest and its Management", the loss of forest areas due to deforestation and transformation into other land-use types, e.g. agricultural area, infrastructure or residential area, can be detected using near-infrared (NIR) or the Normalized Difference Vegetation Index (NDVI). Both, NIR and NDVI generally allow visual conclusions to be reached about the vegetation's distribution, composition, productivity and vitality (cf. Campbell 2002, Lillesand et al. 2004). Spatio-temporal changes in urban areas (module „Land Use") are visualized using imagery with different band combinations, especially involving NIR and mid-infrared (cf. Campbell 2002, Lillesand et al. 2004).

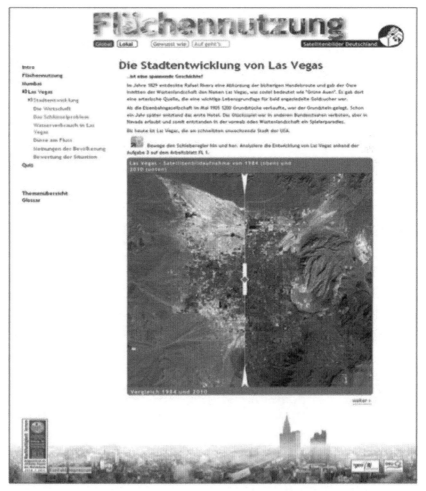

Figure 4. The comparison of satellite imagery of different temporal origin allows users to detect developments or changes within the mapped area in space and time like in the geographic example „City Development of Las Vegas" in the module „Land Use".

Altogether, RS data provide valuable spatial information about a geographic area including local economy, environment and society. In „GLOKAL Change", satellite imagery of different temporal origin and different band combinations is used for visualization purposes. Image interpretation and comparison, in combination with additional information on the topic (texts etc.), are thought to enable the young users to make a well-founded evaluation about the impacts on sustainable development.

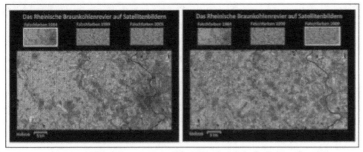

Figure 5. When comparing satellite imagery of the lignite mining area in Western Germany from different time slots, spatial changes in the economic, ecologic and social dimension are observable owing to the mining activity can be observed.

5. USING RS DATA ON A MAP SERVER TO EXPLORE THE INDIVIDUAL HOME AREA

In addition to the satellite imagery provided in the learning modules, „GLO*KAL* Change" comprises a map server containing pre-processed Landsat TM and ETM+ imagery of the entire territory of Germany (Figure 6). Users can access the map server either from the first page of the LP (cf. Figure 2) or from the learning modules. Imagery is available for three time slots, 1985, 2000 and 2007. For each of these slots, one real color and two false color images of Germany, e.g. imagery showing the NDVI, have been processed. Using an overlay function, two images can be viewed at the same time by setting one of them to transparency mode. In this format, the transparent image is on top of the other one, enabling users to compare the images and allowing them to recognize similarities and differences, i.e. changes in the landcape.

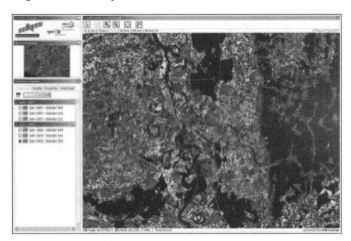

Figure 6. The map server of „GLO*KAL* Change" contains real and false color satellite imagery of Germany for three time slots.

As in Google Earth, users can zoom in/out or navigate on the map server surface using a pan function. Beyond that, users are able to search for certain settlements by name, zip code or geographic coordinates as well as measure distances and areas. Image details that <re interesting for on-site exploration can be printed or downloaded from the map server free of charge, e.g. onto mobile devices (tablet PCs) for application in the field.

Form an educational point of view, the map server is an intermediate step between the geographic examples in the learning modules and an original encounter in the adolescents' local surroundings at home. Thus, overall the perspective shifts from the global- and local-scale examples to the individual local surroundings of the adolescents. This step is thought to increase their motivation and interest in discovering their home area, both on the satellite imagery of the map server as well as in the field.

6. PERFORMING GEOSCIENTIFIC FIELDWORK TO EXPLORE THE INDIVIDUAL HOME AREA

When users deal with the learning modules of „GLO*KAL* Change", they get to know some non-sustainable effects of environmental changes on the economy, environment and society in various geographic areas worldwide and in Germany. In the course of the modules, they also learn to analyze information and evaluate it in terms of their impacts on sustainability (cf. chapter 3). They have to pay attention to all three dimensions owing to the complex interactions between them as a prerequisite to make a well-founded, holistic evaluation. When the adolescents are in the field in the context of an original encounter they require all these abilities to investigate the economic, environmental and social issues in their local surroundings. Coming from the classroom into the field, they have to transfer the knowledge they acquired in the virtual world of the learning modules and using the map server to real situations outside. According to Kirch (1999), students will not really gain geographic knowledge and build comprehension for a geographic area without exposure to fieldwork. Obtaining primary, i.e. non-filtered, information in the field during an original encounter is an important part of the learning process as information from mass media and information systems is filtered by the authors (Haubrich 1997). Furthermore, learners gain individual experience in the field by actively observing or performing fieldwork (cf. Haubrich 1997). Bland et al. (1996, 165) once summarized the overall importance of an original encounter (fieldwork) for learning in geography as follows: „Geography without fieldwork is like science without experiments."

The LP „GLO*KAL* Change" supports adolescents in gaining individual experience and primary information on site by making worksheets available for module-specific, action-orientated geo-scientific fieldwork, e.g. interview guidelines or mapping instructions. For the accurate execution of the fieldwork, „GLO*KAL* Change" also provides papers on background knowledge concerning different geo-scientific fieldwork methods. Using these worksheets and methodology papers the learners can examine the economic, environmental and social situations to some extent, and comment upon their implications, for example regarding the impact of a small gravel pit nearby, or the construction of a road through a forested area.

Once the studetns have studied the international examples and those from Germany in the modules, they can also explore their home area in terms of sustainability. In our view, the personal reference to the home area where the original encounter takes place will be a motivating factor for them. However, they may need assistance on how to obtain meaningful on-site information on the three dimensions. This support is provided by the worksheets and methodology papers. Sometimes, specific information may be needed, which can only be provided using up-to-date aerial imagery. In this case, a micro-drone (see section 6.1) could be applied in the context of „GLO*KAL* Change".

6.1. Using a micro-drone for the generation of specific aerial information

When the learners are in the field they will presumably be able to gather answers to most questions in their geo-scientific fieldwork. However, in some circumstances specific questions may only be answered by using special equipment. During field examinations in the context of „GLO*KAL* Change" a low-flying micro-drone can be used to gather real-time high resolution aerial imagery (Figure 7). As part of the project, the micro-drone can be operated by teachers who take a training course to support adolescents in need of specific information from aerial imagery. The micro-drone's multi-spectral camera can map small areas such as razed forests or damage in cornfields (Thamm & Judex 2005). The micro-drone can be applied to generate information in terms of all four topics presented in „GLO*KAL* Change": open-cast mining of different size, entire forests or small forested areas, agricultural land where energy crops or other crops are grown, and several other uninhabited land-use types, e.g. building sites.

Figure 7. During the fieldwork in the context of „GLO*KAL* Change" a low-flying micro-drone, which is equipped with a multi-spectral camera, can be applied to take real-time high resolution aerial imagery (image on the right).

Beyond the practical benefit of taking aerial imagery for generating information, the micro-drone has an educational function. Before and during its application the adolescents get to know the basics of remote sensing as well as the principles and difficulties of producing RS data. For example, when they learn how an aerial image is taken they may, at the same time, start to understand the process of generating satellite imagery. Thus, the micro-drone is also thought to be an educational learning object. As the youngsters are involved in flight preparation, data acquisition, conditioning and evaluation, they may likely be more motivated and interested in performing fieldwork compared to regular field examinations.

7. CONCLUSIONS AND OUTLOOK

The LP „GLO*KAL* Change" is a multi-dimensional learning environment. By using it in school as well as during extra-curricular environmental education, adolescents can firstly obtain information on economic, ecological and social issues through analyzing and interpreting different types of media including RS data. Secondly, they can get to know examples of non-sustainable development at global and local scales before subsequently examining their own individual local surroundings, a multi-perspective approach. Thirdly, the combination of computer-assisted learning (learning modules and map server) and original encounter (fieldwork) is thought to add to the entire learning process as a multi-sensory approach covering various human sensory channels. Moreover, RS data, fieldwork and the application of a micro-drone are thought to increase adolescents' motivation and their interest to learn effectively in the context of „GLO*KAL* Change". Following the first application of the geographic example „Mining Lignite in the Rhineland, Western Germany" („Mining Resources in Open-Cast Mining") in school, 14 from a total of 22 sixth graders answered with „Yes." while eight said „Maybe again." when they were asked: „Would you like to learn again with GLO*KAL* Change, maybe in terms of another example or topic?".

The overall objective of the LP, which can be used in school as well as in extra-curricular environmental education free of charge, is to foster German adolescents' evaluation and media competence. In the learning modules, adolescents' overall task is to evaluate the impact of various environmental changes on sustainable development. In this context, the comparison and visual analysis of RS data is thought to be an important method in „GLO*KAL* Change" to gain spatial information on economic, ecological and social changes, improving the adolescents' competence to read and interpret satellite imagery simultaneously. Altogether, dealing with a LP like this may bring adolescents one step closer to the aims of ESD, namely (i) to apply knowledge on sustainable development, (ii) be aware of the problems of non-sustainable development, and (iii) act sustainable oneself (cf. section 1), as was intended by the developers.

Thinking of the future, interactive learning environments such as „GLO*KAL* Change" may become increasingly frequent in the study of geographic/geo-scientific content, including content on sustainable development. Haubrich et al. (2007: 248) stated that information and communication technologies (ICT) „… can contribute meaningfully to the aims of education for sustainable development in Geography teaching and learning described in this Declaration [on Geographical Education for sustainable Development] by helping students to acquire knowledge and develop competencies necessary for lifelong learning and active citizenship."

Besides „GLO*KAL* Change", various RS-based online LPs have been developed or are still in development in the Department of Geography at the University of Education Heidelberg, Germany (Ditter et al., accepted), e.g. „BLiF – Blickpunkt Fernerkundung" (www.blif.de). These LPs follow established teaching methods being used in the didactics of geography, general pedagogy and computer sciences. They have been or are being designed to foster the acquisition of knowledge on and individual competences in geography/RS effectively through modern-day computer-assisted learning.

ICT are promoted by the European Union as an important key to improve education and training (cf. European Commission 2010a). The interactive LP „GLO*KAL* Change" belongs to the ICT. A review of several studies of ICT impact on schools has shown „…that ICT impacts on competency development – specifically team work, independent learning and higher order thinking skills…" (ICT Impact Report 2006). „GLO*KAL* Change" aims at fostering several of the adolescents' higher order thinking skills (see section 1) and basically allows users to deal with its learning content either by oneself or in team work. Thus, the LP „GLO*KAL* Change" follows guidelines that have been made by the European Commission to „…develop innovative education and training practices…" (cf. European Commission 2010b). Beyond that, it deals with several sustainability-related topics, which have a European dimension, for example energy supply, use of biofuels and sustainable management of forests.

8. ACKNOWLEDGEMENTS

The authors wish to acknowledge the Deutsche Bundesstiftung Umwelt (DBU) for supporting the project „GLO*KAL* Change – Evaluating global environmental changes locally".

REFERENCES

Bahr, M. 2007. Bildung für eine nachhaltige Entwicklung – ein Handlungsfeld (auch) für den Geographieunterricht?!. Praxis Geographie: 9/2007: 10-12.

Beckel, L. & Winter, R. (eds.) 1989. Satellitenbilder im Unterricht. Einführung und Interpretation. Bonn: Orbit-Verlag Reinhard Maetzel.

Bland, K., Chambers, B., Donert, K. & Thomas, T. 1996. Fieldwork. In Geography Teachers' Handbook, eds. P. Bailey & P. Fox, 165-175. Sheffield: The Geographic Association.

Brucker, A. 2006. Luft- und Satellitenbilder. In Geographie unterrichten lernen. Die neue Didaktik der Geographie konkret, ed. H. Haubrich. 178-179. München: Oldenbourg Verlag.

Bund-Länder-Kommission (BLK) 1998. Bildung für eine nachhaltige Entwicklung – Orientierungsrahmen, Materialien zur Bildungsplanung und zur Forschungsförderung, Heft 69. Bonn.

Campbell, J.B. 2002. Introduction to Remote Sensing, Third Edition. London, New York: The Guilford Press.

de Haan, G. & Gerhold, L. 2008. Bildung für nachhaltige Entwicklung – Bildung für die Zukunft. Einführung in das Schwerpunktthema. Umweltpsychologie: 12 (2): 4-8.

Deutsche Gesellschaft für Geographie (DGfG) 2010. Bildungsstandards im Fach Geographie für den Mittleren Schulabschluss – mit Aufgabenbeispielen, 6. Auflage. Berlin.

Ditter, R., Haspel, M., Jahn, M., Kollar, I., A. Siedmung, Viehrig, K., Volz, D., Siegmund, A.. GeoSpatial Technologies in school – theoretical concept and practical implementation, submitted in: International Journal of Data Mining, Modeling and Management (IJDMMM): FutureGIS: Riding the Wave of a Growing Geospatial Technology Literate Society. In press.

Doering, A. & Veletsianos, G. 2007. Authentic Learning with Geospatial Data: An Investigation of the use of Real-Time Authentic Data with Geospatial Technologies in the K-12 Classroom. eds. C. Crawford, D. Willis, R. Carlsen, I. Gibson, K. McFerrin, J. Price & R. Weber. Proceedings of Society for Information Technology and Teacher Education International Conference 2007. Chesapeake. 2187-2193.

European Commission Webpage (2010a): Strategic framework for education and training. Online http://ec.europa.eu/education/lifelong-learning-policy/doc28_en.htm. Accessed Dec. 2011.

European Schoolnet (2007): The ICT Impact Report. A review of studies of ICT impact on schools in Europe. Written by European Schoolnet in the framework of the European Commission's ICT cluster. Online http://ec.europa.eu/education/pdf/doc254_en.pdf. Accessed Dec. 2011.

Gerber, W. & Reuschenbach, M. 2005. Fernerkundung im Unterricht. Geographie heute: 235: 2-8.

Gross, D. & Friese, H.W. 2000. Geographie, Umwelterziehung und Bildung zur Nachhaltigkeit. Geographie und ihre Didaktik (GuiD): 3/4: 1-44.

Haubrich, H. (ed.) 1997. Didaktik der Geographie konkret. München: Oldenbourg Verlag.

Haubrich, H., Reinfried, S. & Schleicher, Y. 2007. Lucerne Declaration on Geographical Education for Sustainable Development. Published in: S. Reinfried, Y. Schleicher & A. Rempfler (eds.). Geographical Views on Education for Sustainable Development. Proceedings of the Lucerne-Symposium, Switzerland, July 29-31, 2007. Geographiedidaktische Forschungen, Volume 42. Nürnberg. 243-250. Online http://www.igu-cge.luzern.phz.ch/seiten/dokumente/plu_igu_cge_ludeclaration_sustdev.pdf (Accessed Dec. 2011)

Hemmer, M. 2006a. Bildungsstandards im Fach Geographie für den mittleren Schulabschluss (1). Geographie und Schule: 161: 44-46.

Hemmer, M. 2006b. Bildungsstandards im Fach Geographie für den mittleren Schulabschluss (2). Geographie und Schule: 162: 34-40.

Kirch, P. 1999. Vom Kopf auf die Füße. Belebung des Faches Geographie durch Lernen vor Ort. Praxis Geographie: 29 (1): 4-5.

Kollar, I., Wolf, A. & Siegmund, A. 2008. Fostering ‚subjective evaluation faculty' of teenagers in the area of environmental changes by using satellite images in school. In Lernen mit Geoinformationen III, eds. T. Jekel, A. Koller & K. Donert, 70-75. Heidelberg: Wichmann Verlag.

Lillesand, T. Kiefer, R.W., & Chipman, J.W. 2004. Remote Sensing and Image Interpretation. Fifth Edition, International Edition. New York: John Wiley & Sons.

Programm Transfer-21 2007. Orientierungshilfe Bildung für nachhaltige Entwicklung in der Sekundarstufe I. Begründungen, Kompetenzen, Lernangebote. Berlin.

Siegmund, Alexandra 2011. Satellitenbilder im Unterricht – eine Ländervergleichsstudie zur Ableitung fernerkundungsdidaktischer Grundsätze. Dissertation. Fakultät für Natur- und Gesellschaftswissenschaften der Pädagogischen Hochschule Heidelberg. Heidelberg. Online http://nbn-resolving.de/urn:nbn:de:bsz:he76-opus-75244. Dec. 2011.

Siegmund, Alexander & Menz, G. 2005. fernes nah gebracht – Satelliten- und Luftbildeinsatz zur Analyse von Umweltveränderungen im Geographieunterricht. Geographie und Schule: 154 (4): 2-10.

Thamm, H.P. & Judex, M. 2005. Einsatz einer kleinen Drohne für hochaufgelöste Fernerkundung. Eds. J. Strobl, T. Blaschke & G. Griesebner. Angewandte Geoinformatik. Beiträge zum 17. AGITSymposium Salzbug. Hüthig, Heidelberg. 722-730.

United Nations (UN) 1993. United Nations Conference on Environment and Development, Rio de Janeiro, 3-14 June 1992. Volume I Resolutions Adopted by the Conference. New York.

United Nations Educational, Scientific and Cultural Organization (UNESCO) 2011. Education for Sustainable Development (ESD). Paris. Online http://www.unesco.org/new/en/education/themes/leading-the-international-agenda/education-for-sustainable-development/. Accessed Dec. 2011.

REDEFINITION OF THE GREEK ELECTORAL DISTRICTS THROUGH THE APPLICATION OF A REGION-BUILDING ALGORITHM

Yorgos N. PHOTIS
University of Thessaly, Department of Planning and Regional Development, Pedion Areos, 38334, Volos, Greece.
www.prd.uth.gr, yphotis@uth.gr

Abstract

The main purpose of this paper is the formulation of a methodological approach for the definition of homogenous spatial clusters, taking into account both geographical and descriptive characteristics. The proposed methodology, is substantiated by SPiRAL (SPatial Integration and Redistricting ALgorithm), a constrained-based spatial clustering algorithm, whose successive steps focus on the analysis of the characteristics of the areas being integrated, the designation of the spatial clusters and the validity of a joining criterion. We applied the methodological approach and used SPiRAL to solve a realistic electoral redistricting problem. Namely, the redefinition of the electoral districts of the Prefecture of Lakonia in Greece. The results demonstrate an improved layout of the study area's electoral map as far as the problem's criteria and constraints are concerned (adjacency, population and size), justifying in this respect the perspectives and potential of our approach in the analysis and confrontation of similar problems.

Keywords: Spatial clustering, GIS, constraint-based algorithm, electoral districts, Greece.

1. INTRODUCTION

The organization and arrangement of space as it is expressed through administrative, economic and political decisions has critical consequences both at a community and individual level. Geographers, planners and policy makers, during such problem solving processes mainly utilise geo-referenced datasets at a regional or urban scale. The central issue in all of the above cases is the designation of spatially contiguous and robust clusters of areas, which should reflect efficient and competitive regions. Consequently the structure and profile of the newly formed spatial units should stem from a well-structured process complying with predefined criteria and constraints.

The problem of partitioning a territory into districts is widely recognized as a difficult multicriteria, combinatorial optimization problem (Bozkaya et al, 2011). In this framework, the advances in technology and the fast emerging field of Geoinformation have provided considerable impetus for new methodological approaches. Especially in cases, where the geospatial nature of the problem is clearly formulating the solution process, the integration of spatial analysis methods and Geographic Information Systems provides increased the capabilities in data editing, analysis and representation that are needed.

Taking into account the importance of the rational arrangement of space, the aim of this paper is to present a comprehensive methodological approach of spatial clustering. More specifically, when neighboring regions merge to form spatial clusters, whose main characteristics are homogeneity and spatial cohesion, the proposed method deals with both proximity and contingency constraints according to a set of predefined criteria. Underlining the importance of adjacency in the definition of spatial clusters the criteria utilized in the process of redistricting mainly refer to geographical characteristics of regions. The method's successive steps formulate the SPiRAL algorithm (SPatial Integration and Redistricting ALgorithm), which is also described in this paper. The proposed methodology is being applied and tested in the definition of a Greek prefecture's new electoral districts. The criterion set was the homogeneity of population size corresponding to the number of seats.

The paper is organized in five sections through which, the theoretical background and the methodological framework of the approach are presented. The following section introduces a formal definition of spatial clustering problems and a brief overview of methods and techniques applied when dealing with them. The third section focuses on the proposed methodology and the constrained-based algorithm, which during the fourth section are applied to the reorganization of Lakonia's electoral districts. The last section contains some concluding remarks mainly dealing with the performance and the effectiveness of both the proposed methodology and the SPiRAL algorithm.

2. SPATIAL CLUSTERING

Clustering is a key issue in spatial data acquisition and mining since its main purpose is to identify subsets of data with similar characteristics (Grekousis et al, 2012). According to Estivill - Castro, Lee and Murray (2001), "Spatial clustering consists of a partitioning set $P = \{p_1, p_2, ...p_n\}$ of geo-referenced point-data in a two-dimensional study region R, into homogenous sub-sets due to spatial proximity." In this respect, the final definition of spatial clusters is largely depended on the parameters, which specify their number and characteristics such as size, distribution, arrangement and dispersion. During the last two decades, the rapid development and widespread implementation of GIS have triggered a substantiate increase in problems dealing with the formulation of spatial clusters with high levels of homogeneity as their main goal.

One of the major aspects when dealing with spatial planning problems is that the volume of data to be analysed is, usually, large. Spatial data refer to two-dimensional or three- dimensional points, lines and polygons. Exploratory spatial data analysis (ESDA) provides methods and techniques for assisting the processing and exploitation of spatial information, organised in a complex GIS context (Murray et al., 2001). Some of the most applied methods are those dealing with the spatial distribution and organization of data. Bailey and Gatrell (1995) refer to a set of areal proximity measures, such as the distance of centroids, the common border of polygons or the concept of the spatial moving average in order to create a proximity grid. In the unusual case that the distribution of data forms a grid then the median polish technique is applied in order to clarify the spatial trends and tendencies of data. Even though these techniques have been extensively used in geographical research, they cannot by themselves resolve such complex problems. They have to be adopted in a combinatorial framework in order to effectively analyse and explore spatial information according to planning guidelines and objectives.

On the other hand, analytical methods of spatial clustering can be extremely helpful by providing information about the spatial relationships among data (Duque et al, 2007). They aim at the redistricting of space and the best possible allocation of patterns (Openshaw, 1996). In this respect, the interest is focused on finding ways to divide regions in spatial units characterised by similar features.

In recent research and literature a plethora of spatial clustering procedures have been introduced. The most commonly used are spatial clustering algorithms and the main reason for this is their computational accuracy, speed of calculations and robustness of results. According to Sheikholeslami, Chatterjee and Zhang (2000) they can be divided into four broad categories: partitioning algorithms that optimize an equation, hierarchical algorithms which decompose the database, density-based algorithms which exploit a density equation for the determination of the clusters and grid-based algorithms which reduce the dimensions of the clustering space. Alternative point-oriented approaches are provided by the central point method which exploits the distance of artificial points, the median method which is based on the distances throughout the entire point dataset and the based on triplets method which uses the common node of three polygons in order to identify clusters in space (Murray,

1999; Gebhardt, 2000).

As expected, each method is characterized by advantages and disadvantages and thus their exploitation eligibility depends not only on each decision maker's preferences, but also on the study region's geographical profile and attributes. However, in most of the above cases computational demand is extremely high and such methods become time-consuming and burdensome. This critical drawback can be dealt with through the utilisation of Geographic Information Systems whose advanced capabilities in spatial data processing and visualization by maps, graphs, or tables can substantially contribute to their performance and efficiency.

3. METHODOLOGICAL FRAMEWORK

The main aim of the paper is to formulate an alternative and reliable methodology for the formulation of spatial clusters. The proposed methodology should be and is substantiated through a constrained-based algorithm used to organise and regionalize the study area's spatial units in an objective and systematic manner. In such a framework, the analytical fragmentation of the proposed problem solving procedure formulates a constraint-based region-building algorithm. The methodological framework, which confronts the above problem definition, is described in the following paragraphs and depicted in Figure 1.

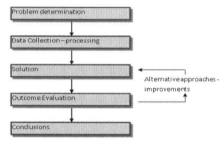

Figure 1. The proposed methodological framework.

Firstly, the problem is defined by its decision variables and constraints determination. The spatial clustering process should succeed in the formulation of homogeneous sub-regions of the study area, which would reflect certain descriptive and geographical characteristics. Since the finally defined regions should preserve spatial continuity and exhibit compact geometry as spatial clusters, adjacency of their structural components is the most important parameter. During the second stage of the process data collection and processing take place. Due to the fact that the central issue in such spatial processes is the topological characteristics of each entity, GIS functions and operations are utilized and problem parameters emerge from their geographical features.

Consequently, the spatial attributes of the polygons such as perimeter, area, metacenter (spatial mean), relative position in space and topological information (i.e. right polygon – left polygon) will be exploited to determine adjacency. Having completed the processing of spatial data, our interest is then focused on the joining method of polygons, in order to schematize the predefined number of clusters. The proposed method should be capable of determining the way that the initial cluster centers are proposed, the geographical variables that will be used for clustering, the distribution pattern that should be followed and also the joining criterion. Such a method should be in an algorithmic form paired with a terminating threshold, whose sequential completion of steps will produce continuous spatial clusters.

In the forth phase of the methodological framework and after the implementation of the algorithm results are examined and evaluated. This stage is critical since new data and modified parameters can be fed to and processed by the algorithm, in order to improve the solution performance and the effectiveness of the approach. Finally, conclusions are derived concerning quantitative metrics of resulting configurations as well as evaluative indicators of the methodology's integrity, spatial redistricting potential and capabilities.

3.1 The Spatial Integration and Redistricting Algorithm (SPiRAL)

As mentioned above, SPiRAL, the proposed constraint-based algorithm combines proximity and contingency measures. In each of its steps the main consideration is formulation of clusters taking into account the spatial distribution of quantitative data, describing the regions (polygons) to be joined. Moreover, the proposed algorithm SPiRAL is consisted of steps and calculations aiming to assist decision makers in clarifying whether the desirable objective is achieved and met. More specifically, SPiRAL evolves as follows:

STEP 1: Definition of the 'join criterion'
The value of the characteristic that the produced clusters should comply to is determined. The join criterion can be either a demographic, an economic or a social parameter and its deviation margins are set and given.

STEP 2: Determination of the n centroids
A - If polygons exist that satisfy the join criterion, they are set as cluster centres.
B - If no polygons satisfy the join criterion, then the following calculations are performed:
1. The maximum (X_{max}, Y_{max}) and the minimum (X_{min}, Y_{min}) value of the spatial average of the remaining polygons is computed.
2. The grid where the **n** centers will be located is defined by calculating each quadrat's side as equal to the division of the difference of the maximum and minimum value of the spatial average for every axe with **n+1** and adding the result (s_1 for axe X and s_2 for axe Y) to the X_{min} and Y_{min} respectively, according to the number of the center. For example, the coordinates of the 1st centre are ($X_{min}+s_1$, $Y_{min}+s_2$), the 2nd ($X_{min}+2s_1$, $Y_{min}+2s_2$) etc.
3. Distances of each polygon's centroid from the every grid point are calculated. The centroid with the minimum distance becomes a center, and polygons satisfying the joining criterion are assigned to the closest grid point, which is then excluded from the set.
4. If adjacent polygons - centers exist, then the polygon with the smaller centroid distance from the corresponding point of the grid, becomes a center, while the other is excluded. Following, the next closest polygon is selected.

STEP 3: Formulation of the initial clusters
Neighboring polygons to each center unit (sharing one or more common borders) are assigned to it.

STEP 4: Allocation of polygons to the initial clusters
A - Polygons adjacent to one and only one center are merged to form a cluster.
B - Polygons adjacent to two or more centers are called meso-polygons and are associated with a cluster following a modified process:
    a) If a cluster meets the joining criterion after the annexation of the remaining neighboring polygons, then the meso-polygon will be assigned to the cluster which does not.
    b) For clusters falling short of the criterion, an Index of Unification I is calculated, which is a standardized form of the distance between two polygon centroids, according to the following equation:

$$I = \frac{d}{l} \quad (1)$$

    where
    d_{jk} = distance between the centers of two polygons j and k and
    l_{jc} = percentage of the common border between a polygon j and a cluster c adjacent to its perimeter.
    The index I is calculated for all of the polygon's neighboring clusters.
    c) The meso-polygon is allocated to the cluster exhibiting the minimum I value.

STEP 5: Threshold criterion
A - Assigned polygons are no further considered.
B - Spatial allocation of remaining polygons continues for each cluster until the join criterion J is satisfied

STEP 6: Allocation of remaining polygons
After the process's completion and the definition of clusters that meet the join criterion J, every unassigned polygon is allocated to its first order neighboring cluster. If a polygon is adjacent to more than one clusters, the Index of Unification *I* is utilised (Step 4: Bb).

STEP 7: Evaluation of the clusters homogeneity according to the join criterion
The standard deviation (σ) for the criterion value is calculated for each cluster according to Equation 2 (Papadimas and Kollias, 1998):

$$\sigma = \sqrt{\frac{1}{N}\sum_{i=1}^{N}(x_i - \mu)^2} \quad (2)$$

where
i = 1,...,n
x_i = the criterion value for each cluster
N = the number of observations equal to the number of clusters, and
μ = the arithmetic mean of the criterion value for the all clusters.

STEP 8: Enhancement of each cluster's quantitative characteristics
A - For clusters not meeting the criterion value, all adjacent polygons are selected.
B - The polygon with the maximum common border percentage with a cluster is assigned to it.
C - Assignment becomes permanent if and only if it improves the standard deviation value with the maximum reduction.
D - If no modifications to the initial polygon set are performed, remaining polygons are processed, according to a descending common border percentage order.
E - Each polygon's adjusted criterion value for each cluster is calculated in order to identify falling short clusters.
F - Adjacent polygons of remaining clusters are located as follows:
   a) For a remaining falling short cluster, an adjacent polygon is a second order common border polygon, if its first order one is already assigned.
   b) For each redefined falling short cluster, an adjacent polygon is one sharing the maximum percentage of common border with it.
G - Any polygon's n-order assignment should not split its n-1-order cluster.
H - If a cluster exists, which can be separated from a polygon improving its standard deviation, then the polygon with which shares the minimum common border percentage is moved.
I - Steps A to F are repeated until every cluster in the study area meets the joining criterion J.
J - Algorithm stops when no further replacement improves the standard deviation value.

The above steps describe the analytical sequence of the spatial clustering process of polygons according to the proposed constrained-based algorithm. The methodology can be applied to a plethora of redistricting problems related, for example, to the delineation of administrative regions, electoral districts as well as educational, emergency response and metropolitan service areas. In the following section of the paper, SPiRAL is utilised in a example for an administrative region's fictitious reorganisation while on the same time, its efficiency is tested and evaluated.

4. REDEFINITION OF LAKONIA'S ELECTORAL DISTRICTS, GREECE.

Greek electoral districts, not surprisingly characterised by geodemographic and socioeconomic disparities, also seem to be defined by random boundaries that form unadjusted shapes (Valasaki and

Photis, 2005). In this respect, they constitute an intriguing and challenging study area to which, the proposed methodological frameworks as well as SPiRAL were applied in order to redesign one of its existing electoral districts namely, the Prefecture of Lakonia and formulate new homogenous single-seat units by clustering contiguous municipalities.

4.1 Study Area and Data

The current electoral districts of Greece were established by the unification of a number of municipalities and communes (declared as spatial units after the implementation of law 2539/97 for the 'Reformation of the 1st Degree of the Local Administration'). Their borders are consistent to the latter and the number of seats varies with relevance to their population size. Consequently and with respect to problem formulation, structural units are municipalities and communes, whereas reference unit is the electoral district. All data utilized throughout the solving process are geo-referenced, mainly concern geographical and descriptive characteristics of polygons and were provided by the Laboratory for Spatial Analysis GIS and Thematic Cartography at the Department of Planning and Regional Development in the University of Thessaly. The join criterion is defined as the equal distribution of the district's population per seat and the respective demographic data relate to the 2001 Census of the National Statistics Service of Greece.

Regarding processing stages and environments, the proposed methodology was basically realised in ESRI's ArcGIS platform. Polygons centroids were selected and the topological data of lines and polygons of the region were calculated. Editing also includes calculations concerning the line percentage of common border of every polygon's perimeter. In a similar manner, contingency tables were created for the entire set of structural polygons where the adjacency percentage of common border with their first order neighbours was added as an attribute.

4.2 Application

According to the successive steps of SpiRAL, its application starts by determining the join criterion to which spatial clusters should comply. To this end, with the 2001 Census stating that the total population of Greece is 10.259.900, with nearly two thirds living in urban areas and since the number of parliament seats is 288, the join criterion, representing the electoral standard of the country, is 35.624 people. The acceptable deviation, in this case, is set to 30%, therefore the population of any one-seated electoral district may vary from 24.936 to 46.312 people. If necessary, different deviation margins can be set by the decision makers. Lakonia's electoral district is located in South Continental Greece, with a population 95.696 people and three (3) respective parliament seats. First, the centres of the initial clusters are defined according to the polygons spatial features. Populations of the district's municipalities show that there is no structural unit satisfying the join criterion. Consequently, centres are defined with respect to each polygon's centroid (Step 2B). Taking into account the restrictions that the algorithm sets concerning adjacency, as initial centres are set the municipalities of Githio, Skalas and Niaton (Figure 2).

In the next stage, the remaining polygons are initially allocated to the three centers according to Steps 3 and 4 of the algorithm and simultaneously processed for all districts. More specifically, Anatoliki Mani, Oitilou and Smynous will be assigned to the Githio cluster, whereas for the municipality of Krokeon the index I will be calculated since it has common borders with Skala. Similarly, the municipality of Therapnon will be unionised to Skala, while for the remaining two units (Elous and Geronyron) adjacent to the second center, the index I will determine which of the two clusters (Skalas or Niaton)they will join. Lakonia unionises with the municipalities of Zaraka and Molaon. Calculated values of the index I indicate that Geronyron and Krokeon should join Skalas cluster, whereas Elous that of Niaton. As a result, all municipalities are assigned to the three clusters until, according to the termination condition of the Step 5, the join criterion is satisfied.

After the allocation of the remaining polygons (Step 6) the three clusters are initially defined and the up to this point (common border and distance of the centroids) electoral map of Lakonia is depicted in Figure 3. The synthesis as well as the total population of the three defined clusters is shown in Table 1.

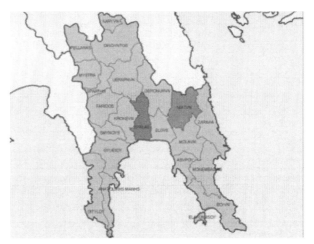

Figure 2. Initial cluster centers.

A first and obvious conclusion deriving from the table is that the district population distribution needs to be optimised in order to achieve the desirable level of homogeneity between the newly designed clusters. Improvement of the clusters quantitative characteristics is accomplished by means of standard deviation of population per seat calibration according to Steps 7 and 8.

The standard deviation of the three clusters of Lakonia is 10.981. Potential reallocations can be considered for the polygons of Krokeon and Spartis to the cluster of Githio that is the only one meeting the defined electoral standard. In this respect, transferring the Krokeon municipality is initially examined since it shares the longest common border with the specific cluster. In this case, the standard deviation drops to 8.110, therefore the movement is made permanent and the population of the cluster rises to 23.808 people. When the Spartis polygon is transferred standard deviation increases to 8.212 and thus joining is not applied.

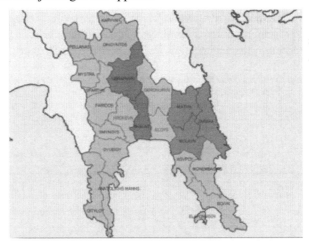

Figure 3. Polygon allocation according to adjacency.

The population of the three clusters that have just formed varies as the following Table 1 shows.

Code	Municipality	Population (2001)	Total Population
1613	NIATON	2557	31860
1607	ZARAKA	1696	
1610	MOLAON	5472	
1606	ELOUS	5992	
1611	MONEMBASIAS	3950	
1602	ASOPOU	3666	
1603	VOION	7802	
1621	ELAFONISOU	725	
1617	SKALAS	6919	42899
1608	THERAPNON	2999	
1604	GERONYRON	2034	
1609	KROKEON	2871	
1614	OINOUNTOS	2649	
1619	SPARTIS	16322	
1622	KARION	660	
1616	PELLANAS	3863	
1612	MYSTRA	4582	
1605	GITHIOU	7542	20937
1618	SMYNOUS	1537	
1615	OITILOY	4985	
1601	ANATOLIKIS MANIS	2024	
1620	FARIDOS	4849	

Table 1. Cluster composition and populations.

Since there are no possible movements left to the cluster of Githio, the cluster with the maximum population (Skalas) is examined whether one of its polygon units can be separated. Starting from peripheral polygons, in ascending order, those of Skalas, Geronyron and Spartis are selected. Their staged examination, results to the reallocation of Skalas and Geronyron to the cluster of Githio, decreasing standard deviation to 844 people. The finally resulting area scheme according to our methodology is depicted in Figure 4 while the respective populations of the three redefined electoral districts are shown in Table 2.

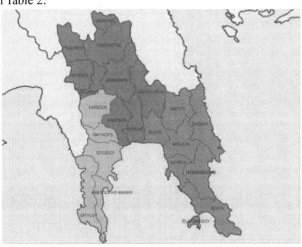

Figure 4. New electoral districts of Lakonia.

Code	Municipality	Population (2001)	Total Population
1613	NIATON	2557	31860
1607	ZARAKA	1696	
1610	MOLAON	5472	
1606	ELOUS	5992	
1611	MONEMBASIAS	3950	
1602	ASOPOU	3666	
1603	VOION	7802	
1621	ELAFONISOU	725	
1608	THERAPNON	2999	31075
1614	OINOUNTOS	2649	
1619	SPARTIS	16322	
1622	KARION	660	
1616	PELLANAS	3863	
1612	MYSTRA	4582	
1605	GITHIOU	7542	32761
1618	SMYNOUS	1537	
1615	OITILOU	4985	
1601	ANATOLIKIS MANIS	2024	
1609	KROKEON	2871	
1617	SKALAS	6919	
1604	GERONYRON	2034	
1620	FARIDOS	4849	

Table 2. Population size of the new electoral districts of Lakonia.

5. CONCLUDING REMARKS

The need for objective and unbiased organization of space, based on a set of predefined quantitative criteria, is intense. Spatial clustering of administrative regions is a critical planning issue with an extremely wide field of applications. At the same time is politically sensitive in the sense that it can be applied in favour or against one or more regions. The proposed methodology as it was realised by the SPiRAL algorithm, can improve the geographical compactness of a region through simple and comprehensive steps and towards the goal of homogeneity.

Both the conceptual spatial clustering framework and the overall regionalization approach allow the simultaneous consideration of variables that refer to and ensure the adjacency and homogeneity of constitutional units. Furthermore, SPiRAL algorithm provides the ability to optimize the outcome by analysing and smoothing the criterion-related deviation of the final clusters. In this respect, the question "how many regions" is effectively and alternatively confronted under specific constraints and requirements.

Considering the algorithm's spatial nature, the exploitation of GIS during the execution of the algorithm, is essential not only due to the topological information processed, but also to their increased spatial analysis and cartographic representation functionality. The extended spatial databases needed remain a critical problem parameter that significantly contributes to some time-consuming issues of the approach. Improvements with respect to the user interface and the integration of the SPiRAL algorithm's steps to a unified open-source system platform should be considered as a fundamental prerequisite for its further utilisation. According to the literature, as GIS become more involved in spatial problem solving procedures, the overall process is significantly improved through the fluent analysis of a set of hypotheses and scenarios and the definition of solutions exhibiting advanced levels of effectiveness and efficiency.

With respect to the algorithm, an intervention that will improve the region-building outcome is the consideration of the second order contiguity in the determination of the initial clusters and the calculation of the index of Unification I. In this manner, more complicated spatial problems can be resolved since the analysis of space will be more thorough. In a similar framework, more than one criterion can be exploited and interrelated (service area size, average distance travelled and structural capacity to name a few) widening its spectrum of applications. As a result decreased processing times will be achieved as well as meaningful and useful regions will be obtained.

In conclusion, SPiRAL algorithm succeeded in defining homogenous spatial clusters according to a specific criterion with the resulting regions meeting the set constraint. The fictitious case study underlined the importance of adjacency in the determination of administrative regions while reassuring the rational formulation of spatial patterns. However, in using methods like the one presented, decision makers need to be aware of several practical issues that stem from and are reflected to the input data variable(s). Their carefull and justified selection is not only essential for but a truly sine qua non condition.

REFERENCES

Bailey T.C., Gatrell A.C., (1995) Interactive Spatial Data Analysis, Essex: Longman Group Limited.

Bozkaya B., Erkut E., Haight D., and G. Laporte. 2011. Designing New Electoral Districts for the City of Edmonton. Interfaces, 41, 6.

Duque J. C., Ramos R., Suriñach J., (2007) 'Supervised Regionalization Methods: A Survey', International Regional Science Review, 30, 195.

Estivill-Castro V., Lee I., Murray A.T., (2001) 'Criteria on Proximity Graphs for Boundary Extraction and Spatial Clustering', PAKDD 2001: 348-357.

Gebhardt F., (2000) 'Spatial Cluster Test Based on Triplets of Districts', Computers and Geosciences, 27: 279-288.

Grekousis G., Photis Y. N., (2012), A fuzzy index for detecting spatiotemporal outliers, Geoinformatica, Vol 16, No 3.

Grekousis G., Photis Y. N., Manetos P., (2012), Modeling urban evolution using neural networks, fuzzy logic and GIS: The case of the Athens metropolitan area, Cities (in press).

Koutelekos J., Photis Y. N. and P. Manetos (2007), Geographic Information Analysis and Health Infrastructure, Health Science Journal, Vol. 1, Issue 3.

Koutelekos J., Photis Y. N., Milaka K., Bessa N. and P. Manetos, (2008), Primary care clinic location decision making and spatial accessibility for the Region of Thessaly, Health Science Journal, Vol. 2, Issue 1.

Murray A. T., McGuffog I., Western J. S., Mullins P, (2001) 'Exploratory Spatial Data Analysis Techniques for Examining Urban Crime. Implications for Evaluating Treatment' The British Journal of Criminology 41:309-329.

Murray A.T., (1999) 'Spatial Analysis using Clustering Methods: Evaluating Central Point and Median Approaches', Journal of Geographical Systems, 1.

Openshaw S., (1996) 'Developing GIS-Relevant Zone-Based Spatial Analysis Methods', στο Longley P., Batty M. (ed) Spatial Analysis: Modelling in a GIS Environment, Cambridge: GeoInformation International.

Papadimas O., Koilias C., (1998) Applied Statistics, Athens: New Technology Publications. Ricca, F. and Simeone, B., (2008), „Local search algorithms for political districting,"
European Journal Operations Research, v189.

Sheikholeslami G., Chatterjee S., Zhang A., (2000) 'WaveCluster: a wavelet-based clustering approach for spatial data in very large databases', The VLDB Journal, 8.

Valasaki M.K., Photis Y. N. (2005) 'Convergence and Divergence of the Greek Electoral Districts: an Approach with the Exploitation of the Quantitative Spatial Analysis Methods', Technika Chronika, 1.

MEASURING EQUITY AND SOCIAL SUSTAINABILITY THROUGH ACCESSIBILITY TO PUBLIC SERVICES BY PUBLIC TRANSPORT. THE CASE OF THE METROPOLITAN AREA OF VALENCIA (SPAIN)

María-Dolores PITARCH GARRIDO
Instituto Interuniversitario de Desarrollo Local Departamento de Geografía Universitat de València,
Av. Blasco Ibáñez, 28, 46010 Valencia SPAIN
http://www.uv.es/uvweb/departament_geografia/en/departament-geografia-1285858446156.html,
maria.pitarch@uv.es

Abstract

Spatial equity in complex spaces such as metropolitan areas is a very interesting subject for research, particularly in view of its enormous potential public policy applicability. An approach to the subject based on the population's access to essential public services (education, health care and social services) is proposed. Geographic Information System (GIS) tools have made a powerful contribution to the ease with which both spatial and statistical data can be handled. The study covers the Metropolitan Area of Valencia, in Spain. It is based on the location of public facilities and the population's ability to move around using public transport. Its objective is to give a general overview of the situation and point to problem zones, with the aim of suggesting answers for these that could help to improve social and spatial sustainability and equity in this metropolitan area.

Keywords: *Spatial equity, accessibility, public services. Valencia Metropolitan Area, social sustainability, mobility*

1. INTRODUCTION

The study reported in this paper aimed to measure equity in the population's access to basic public services, based on their accessibility to the population as a whole, by means of a GIS (Geographic Information System). The study area is a Spanish metropolitan area, in this case Valencia. Its inherent spatial complexity makes it a laboratory-territory of the utmost interest for verifying the results at local level. The applied part of this study comprises two major sections: the development of a method for measuring accessibility and the application of the method, based on constructing a GIS. The latter was highly complex, largely because of the lack of information on the provision of public services in the area, or its inaccuracy, and because there was no map of public transport networks, which had to be drawn up in order to carry out this survey. The result is a very powerful analytical tool, not only for the study presented in this paper but also for future territorial studies.

The concept of accessibility has an interesting dual dimension, being both geographical and social. The present paper focuses on the former as an indirect measurement of the latter. In other words, the measurement of physical or geographical accessibility contributes to the knowledge of whether public services are being provided adequately – equitably – to serve the whole population of the study territory irrespective of where people live.

Many authors have published papers and studies, both theoretical and empirical, on the subject of accessibility. There are many different indicators of acceptability, mostly based on distance and user satisfaction. The main difficulty resides in the measurement process itself, more because of the quantity of data that need to be handled than because of the formulation of the indicators. The use of

GIS has helped to make this task easier and, consequently, to expand the possibilities of present and future analyses.

In the more advanced countries, the welfare state sees to providing the population with universal basic services. This has become an inalienable social right that guarantees equal opportunities for every group of citizens and reduces marginalisation and poverty by guaranteeing unpaid access to these services. New social demands have arisen that call for better quality services, including the need for them to be suitably located: this is considered one of the main conditions for their meeting the minimum requirements of both equity and efficient public investment. The dynamism of society, and all the more so in complex spaces with multiple living interrelations such as metropolitan areas, involves changes in the location of homes and workplaces, rises and falls in the populations of certain areas, etc. This also entails changes in access to services unless the services are adjusted, relocated or adapted to the changes so that the minimum threshold of equity is never lost. GIS makes it possible to carry out this analysis continuously by updating the information, making it a very powerful tool, as shown in this paper.

Of the different types of service provided, this study centres on ones that a number of authors consider fixed (the user goes to the service) and with free universal access (Mérenne-Shoymaket, 1996; Calvo et al., 2001). These are public services that are required by the entirety of the potential demand and therefore need to be located in spaces with good access. It should also be pointed out that public services, together with community facilities and communication infrastructure, are currently one of the motors of local and regional development that make a clear contribution to territorial rebalancing and, consequently, to meeting social equality and equity criteria, fully justifying studies such as this which are of great practical use for decision-makers.

The present study centres on a tangible, measurable aspect, accessibility based on public transport (city bus, metropolitan area bus, the underground/tram network and the local railway network), which has clear repercussions for social and territorial equity (based on the location of public service provision and of the population), with reference to a specific area, the metropolitan area of Valencia (Spain), which covers nearly 500 km^2 and has a population of over 1.8 million. These three factors give a closer view of local realities that may provide a useful example for the study of other similar urban areas, although this study is also of interest in itself, since an examination of territorial equity based on access to public services is one of the components of basic preliminary studies for local land use plans, which are still non-existent in many metropolitan settings.

Accessibility is one of the possible measures of social sustainability, as mentioned above, but it is not the only one. The public transport network is not the only way to reach public services, either, nor are these all of the same type, nor do consumers all show the same behaviour. However, the combination of public services and public transport gives an initial picture of the degree of equity in the study area that can serve as a starting point for subsequent, more detailed studies.

2. PUBLIC SERVICES AND SPATIAL EQUITY

The subject of spatial equity and how to measure it through accessibility has seen little variation since the 1970s (Garner, 1971; Harvey, 1973; Domanski, 1979). The main research aim continues to be concerned with how to achieve greater spatial equity without necessarily sacrificing a degree of economic efficiency. Accessibility is a frequent basis for models and explanations that attempt to throw some light on the implications of the geographic location of public services. Different types of model have been proposed, ranging from highly centralised ones to others that favour extreme dispersion, and from the most theoretical to the extremely applied, but what almost all of them have in common is maximising the population served by government-backed public facilities, programmes or action. One way to achieve this objective is to improve their access by public transport.

Harvey (1973) was one of the first geographers to define the term spatial equity, also known as spatial justice. Spatial justice must pursue the following aims: respond to the needs of people in each territory, assign resources to maximise spatial multiplier effects and assign extra resources to help overcome the problems that have their origin in the physical and social environment. Spatial justice depends on accessibility and on other factors such as supply quantity, the degree of availability of the services, etc. Both efficiency and spatial equity are particularly relevant for public services, as has been pointed out.

The economic activity location models that have been developed since the 1950s, particularly those for public services, attempt to find an optimum location to achieve the maximum return on the supply. However, the reality is often far more complex than the models take into account. Political factors associated with local decision-making or with very different public priorities have created a network of public provision of the main welfare services (health, education and social services) that does not always respond to this optimum location. Traditionally, the standard measurement tool has been the ratio of variation of demand inputs (such as pupils per teacher, doctors per thousand inhabitants, etc.). However, this bears little relation to measurements of accessibility, which clearly contribute to measuring the efficiency and equity of the location of public services. The balance between two factors that roughly speaking can be called size and distance helps to define equity in access to the service and efficiency in its use, in that it can serve a particular demand. This subject is particularly relevant in urban and metropolitan areas with high population concentrations in particular spaces and considerable dispersal in others.

Nowadays, studies are adopting a more practical bent as an aid to decision-making. The location of services already exists and is difficult to change, although it can always be improved. The best location for a service does not always entail moving it: on occasion, as already mentioned, better access would be the answer. Improving the transport network and/or setting up new networks is essential nowadays to integrate and organise urban and metropolitan areas, where the spread and complexity of urban development are inevitable. What is known as smart urban growth takes sustainability as the basis for urban planning, but its bias towards managing growth and environmental aspects would seem to sideline somewhat the problems of social equity (Foster-Bey, 2002).

Accessibility is a basic geographical concept. Equitable accessibility is a complex matter (Crooks and Andrews, 2009). It is related to many questions such as decisions about assigning resources, the location of the service or activity, information, or even the quality of the service. In short, it means how „easily" a user can obtain the service that he or she needs. For this, physical accessibility is important but so is its measurement in terms of time, since as Miralles (2011) pointed out, the social times (mobility times) of the city „draw the everyday spaces of the metropolitan regions" (p. 127). Travelling time contributes enormously to the citizens' view of the quality of the public services provided and, therefore, the quality of their everyday life. Time is a measurement that relates activities to places (May and Thrift, 2001; Davovidi, 2009). This refers to social time, which brings together a spatial variable, related to the location of the activities in the territory, and a time variable, the result of the time spent on everyday activities, including journeys. The social use of time is, therefore, closely related to the use of the city and of the metropolitan space. The physical makeup of this space and of the infrastructure supporting mobility strongly influences every type of territorial dynamics and makes a powerful contribution to defining the quality of life of its citizens (Mückenberg, 2009; Miralles, 2011). Proximity is an increasingly valued aspect of a territory's quality and of social welfare.

Moreover, one of the problems most frequently studied and condemned by social scientists in many countries is that the social structure of cities and, particularly, of metropolitan areas, is undergoing major changes that are not being paid the attention they deserve. In Spain, for instance, urban planning is based almost exclusively on reference to the land, its ownership and its price, obliging the citizens to adapt to the city rather than the other way round, which is what should happen. Spanish urban planning has an effect on social cohesion or disintegration (Bueno Abad and Pérez Cosín, 2008).

It is true that much progress has been made on the subject of the impact of accessibility on the equitable provision of services, but the question of optimum travelling distances is still not clear. Schuurman et al. (2010) suggest that the term ‚optimum' is best used when comparing methods rather than for seeking or modifying spatial accessibility. The key lies in the process of interpreting the results so that they will be useful in a possible political decision-making process.

3. SUSTAINABILITY AND EQUITY IN METROPOLITAN AREAS

Metropolitan areas are complex spaces in which interactions between the different territorial processes that take place in them are both the cause and the consequence of how they are organized, how they have evolved, and where infrastructure, services, residential spaces, industrial spaces, etc. are located. Metropolitan areas are becoming especially relevant because they are territories that have traditionally been made up of a central city and its hinterland but are now organised in a complex way, with multiple peripheral situations and new centres within the periphery which are fostered by mobility and not always by proximity to the centre (Corral, 1994).

Metropolitan areas in Europe have seen enormous growth over the past five years, as have urban areas in general, leading the European Union to develop various policies to help manage the different problems found in them. The Leipzig Charter on Sustainable European Cities adopted by the European ministers recommends that Member States pay particular attention to the growth and planning of urban spaces from an integrated and sustainable perspective, particularly in the most deprived neighbourhoods. In the EU's recent cohesion policy (the 2007-2013 plan) the urban dimension has been brought fully into the programmes and projects co-financed by the European Regional Development Fund (ERDF), meaning that integral development was set in motion in these areas, both horizontally and vertically, with greater responsibilities and investments devolved to the local level in response to the growing complexity of these territories (European Commission, 2009).

In Spain, a pronounced decentralisation process is taking place within the metropolitan areas but the cities at their centre are not losing their influence. The population shift from the central areas to the metropolitan rings has also joined that from the most populated and densest nuclei to medium-sized and smaller ones, generally with a low population density (Nel.lo, 2001). The structure of metropolitan space is closely linked to transport infrastructure, which is both the consequence of and a contributor to the suburbanisaton process and to a shift from monocentric to polycentric structures in the internal organisation of the transport infrastructure itself. This is due to changes in the basic patterns of accessibility, which also explains the strength of the demand for journeys and their modal distribution (Schwanen, Dieleman and Dijst, 2001; Albertos et al., 2007; Gutiérrez and García, 2007; García, 2010).

All this tells us that, at the very least, spatial equity has varied over the course of this process. The changes in accessibility induced by new road infrastructure, cause and consequence of the suburbanisation process, create new inequalities for the population in both the old and the new areas of development.

Public services are not equally accessible everywhere, in other words, space introduces some forms of exclusion. In their complexity, metropolitan areas present imbalances that can, on occasion, be particularly striking. In an attempt to reduce these exclusions to as few situations as possible and achieve a fairer spatial distribution, some localisation models have been developed with criteria such as public utility (the number of people who use the service) or travel costs.

The latter notion is the basis of the nodal or functional region concept and is also fundamental for mobility models and to explain the spacing of certain activities. Whatever the activity, but particularly if it involves services to the population, its area of influence extends beyond the exact spot where it is located. Since these centres are spatially at a distance and their services are mainly provided face to face – in other words, the user has to travel to the place where the service is located – connections between them are essential and one of the basic premises for studying them is to consider them nodes or focal points of the transport network.

The quest for social equity, together with territorial equity in metropolitan areas, is key to achieving sustainable territories, with all what that implies in terms of improving the inhabitants' quality of life. From this perspective, the current economic climate (which involves greater competition between territories and less availability of public resources) entails, among other things, a greater need to manage local resources efficiently, to lead the shared effort of local bodies and organisations to pursue more sustainable development and to introduce innovative management models that will make it possible to improve the quality of life of the population. Decision-makers must increasingly have strategies to make local government policies easier for industry, the unions and the public in their areas to understand. The EU has therefore been using different strategies to strengthen this aspect, such as the EU Territorial Agenda or the EESC Opinion on European metropolitan areas (OJEU C 168/10, 20.7.2007).

4. METHOD: THE SPATIAL SEPARATION INDEX

In this accessibility analysis the first data obtained were the distances between the basic territorial unit for which census information is available, the census tract (taking its built-up centre as the starting point for any journey), and the exact point where the facilities of each of the three types of basic service considered in this study are located. The distances were calculated in time (minutes), as this is this measurement that determines user satisfaction and gives a better comparison of the efficiency of each of the possible public transport modes. The three types of service considered were health care, education and social services. Within the first type, a distinction was made between hospitals and health centres; within the second type the distinction was between primary and secondary schools; of the social services, only the basic ones were considered. In every case strictly private facilities were excluded and only public and subsidised services were taken into account.

The model employed involves calculating urban mobility over the networks on which it takes place, in other words, the aim is as real a model as possible. To achieve this, TRANSCAD 6.0 transportation GIS software was used. Three types of transport or mobility networks were studied: the pedestrian network (it will always be necessary to make at least part of the journey on foot), the city bus and metropolitan bus network, the underground/tram network and local trains. In all cases, the transport service frequencies and their operating speeds were considered for all the different lines. In this way it was possible for the distance calculations to include the time taken to get to the public transport, wait for it and change between transport modes as well as the actual travelling time. In other words, the journey time calculation is door-to-door, from the centroid of the census tract to the exact location of the service in question.

Out of the wide range of existing indices (Garrocho y Campos, 2006; Bhat et al. 2000), the Spatial Separation Index was chosen. This calculates the mean distance in minutes between two points. It is simple and easy to interpret. In this type of index, all the starting points carry the same weighting in the calculations and the index only reflects the data on distances. Because it is so simple, it can be used to compare different situations (such as access to public and to private services or to different classes of service) clearly and efficiently. Complexity was introduced by using the real, verified times of the real mobility network, making these results very reliable.

Accordingly, the Spatial Separation Index (Índice de Separacion Espacial) for spatial unit i (ISE_i) is:

$$ISE_i = \sum_{j=1}^{n} \frac{D_{ij}}{n}$$

where

i is the basic spatial unit (census tract) for which the index is calculated, which is taken as the possible starting point for a journey

j is each of the possible journey destinations (the services)

D_{ij} is the distance in minutes between the starting point i and the destination j, based on the matrices calculated and

n is the number of possible destinations.

The calculation only took into account the basic public service nearest to the census tract where the population lives. This analysis therefore assumes that the citizen will travel to the public service closest to his or her home.

5. THE METROPOLITAN AREA OF VALENCIA

The case analysed here is the Metropolitan Area of Valencia (MAV), on the east coast of Spain, south of Barcelona and east of Madrid (Map 1). Industry and the service sector predominate in this area, which has good communications. It is part of the Autonomous Region of Valencia, the fourth most-populated in Spain, with 5,011,548 inhabitants at 1 January 2012. The region's main cities are Valencia, Alicante and Castellón, but only Valencia has a true metropolitan area, the largest in Spain after Madrid and Barcelona, which accounts for 37% of the population of the region and 75% of that of its province.

The MAV revolves around a central city, Valencia (population 798,033 in 2011), and 75 municipalities with slightly over one million inhabitants within a radius of almost 40 kilometres, taking the total to 1,862,053 inhabitants. It is a complex area in terms of the dispersion of its population and built-up areas. Over the 2001/20011 period its population increased by 261,255. Since the city of Valencia only grew by slightly over 50,000 inhabitants over the decade, a real mean annual growth rate of under 1% and less than the 1.67% of the metropolitan area as a whole, this growth mainly took place in medium-sized settlements and, to a lesser extent, in the larger centres with better communications (Torrent, Sagunto, Paterna, Mislata, etc.).

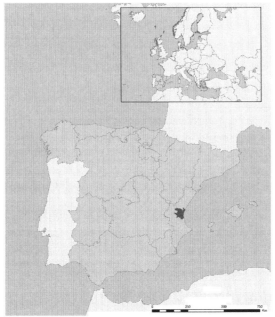

Map 1. Location of the study area

In the past twenty years the MAV has undergone a major urbanisation process that has increased the density of both its built space and its population (Table 1). At the same time, new communications infrastructure has been put in place, particularly the building of the underground and the extension of bus lines within already urbanised spaces and to the newly built-up areas (Map 3). The expansion of the road network for private vehicles has been fundamental for the spread of an extensive urban development model based on family houses strongly linked to open natural spaces, not forgetting the development of inner city spaces and expansion of their edges through the mushrooming of comprehensive action plans (Plan de Acción Integrada – PAI), used as a way to modify the general town plan (Plan General de Ordenación Urbana – PGOU) and reclassify agricultural land as urban or building land. Although it is also a result of space infilling, the combination of three factors as both cause and, in turn, consequence, namely demographic growth, expansive urbanisation and the building of a wider network of communications infrastructure, explains the consolidation of a metropolitan structure in which the zone that is furthest away from the central city is precisely the one that has seen the greatest increase in population density, indicating a relatively greater population growth and urban development expansion towards the periphery, which offers advantages like lower land prices and closeness to natural spaces (Graph 1).

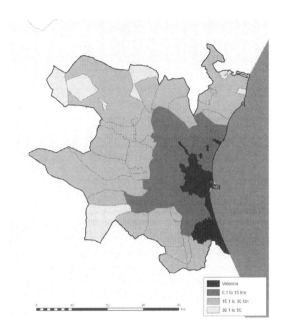

Map 2. Metropolitan rings according to distance from the centre of Valencia city

Scope	Distance from Valencia	Surface area (km²)	Population density 1996 (inhabs/km²)	Population density 2000 (inhabs/km²)	Population density 2011 (inhabs/km²)
Valencia	0	134.63	5546.19	5489.22	5927.60
First ring	1-15 km	456.42	1250.91	1293.72	1591.14
Second ring	15.1-30 km	1317.71	168.95	176.72	235.54
Third ring	>30 km	277	67.84	69.90	98.98
Total MAV	-	2185.8	714.24	723.86	851.89

Table 1. Population and density by metropolitan ring (Source: Own elaboration)

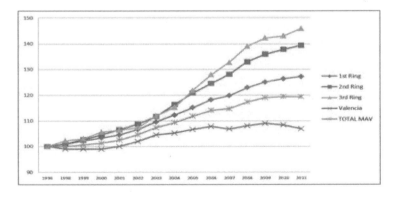

Graph 1. Evolution of population density by metropolitan ring

This territorial restructuring has not always been accompanied by an improvement in the provision of public services and facilities, particularly as regards public transport and the services in greatest demand: health, education and social services. The urban expansion has therefore generated considerable private mobility, compounded by the proliferation of shopping and leisure centres on the periphery of the MAV in recent decades. The public transport network is particularly dense in the central area and the first ring, but sparse or non-existent in the second and third metropolitan rings. These are precisely where the most intense growth is taking place, as mentioned above, and since the public transport network and public services tend to be located in the central zones, the outer rings are also where the greatest inequalities in access to these services are found. The result is considerable private mobility, with the sustainability problems it entails, or growing inequalities in access to services which are accentuated in particular population groups such as old and young people, who have more limited access to private transport. The problems of spatial equity are real and a complex territory such as the MAV is an interesting laboratory for testing some of the ways of measuring the imbalances and inequities introduced by the space and its characteristics.

6. EQUITY IN ACCESS TO PUBLIC SERVICES BY PUBLIC TRANSPORT IN THE VALENCIA METROPOLITAN AREA

6.1. Overall Results

Calculating the index of accessibility by public transport in the MAV has produced some interesting results. Although public transport is not the most efficient in terms of journey time, what is of interest is its public nature, which in principle makes it accessible to the entire population and enables the equity of a territory to be measured. Generally speaking, because the provision of transport in the metropolitan core and first ring is greater than in the periphery, the further away from the central city the longer the travel time by public transport. The MAV shows a two way process: on the one hand decentralisation of activities and residence, and with them the provision of services, and on the other hand, intensification of the most immediately local space in the main city and town centres.

The Spatial Separation Index calculated in this study showed significant differences by type of service (Table 2). The best access was clearly to primary schools (mean ISE = 7.12), as this is a lower (and therefore better) value than those for the social services (mean ISE = 13.85) and primary health care (mean ISE = 13.88), always bearing in mind that these figures refer to public not private services reached by public transport alone. For comparison, the same index was calculated for private transport. The journey times were better in every case and very significantly so in the case of hospital accessibility (ISE = 10.38). As this is a more specialised level of health care, hospitals are fewer in number and dispersed over the area, so those living in the most distant zones have a longer journey time if they use public transport because they need to change buses or trains, involving waiting times which the use of private transport avoids.

The services that were most evenly and equitably distributed over the metropolitan area in relation to the population were primary schools (some of which are also secondary schools). They form a dense network in which closeness to the demand is a priority and constitute the nucleus of service provision. The fact that a large number of them are subsidised rather than exclusively public explains their dispersion over the territory and the very low mean journey times. The same is true of secondary schools, which also present low journey times. Social services and health centres were in an intermediate position, with good territorial distribution, in keeping with the fact of their often being neighbourhood services. Also, in the case of the social services, political initiatives by many town halls have been decisive in improving their spatial distribution. Lastly, as is only natural, the worst accessibility is to public hospitals, of which the MAV has nine.

Map 3. Public transport network in the MAV

EMT: city buses; Renfe-Cercanías: local trains; Metrovalencia: underground/trams; Metrobus: Metropolitan area buses

Service	Mean ISE in minutes (simple)		Mean ISE (weighted by population)		Number of centres provided
	Public transport	Private transport	Public transport	Private transport	
Hospitals	34.80	10.38	37.14	11.02	9
Health Centres	13.88	4.34	14.65	4.55	77
Basic social services	13.85	4.08	14.21	4.17	98
Primary schools (public and subsidised)	7.12	2.77	7.62	2.92	447
Secondary schools (public and subsidised)	9.94	3.40	10.65	3.60	263

Table 2. Spatial Separation Index for the Valencia Metropolitan Area (Source: Own elaboration)

The above analysis was complemented by considering the population involved (Tables 3 and 4). The health and social services present asymmetrical distribution. In these two cases the highest percentage of the metropolitan population has middling accessibility (between 10 and 30 minutes), whereas for education services the highest percentage of the population is under 10 minutes away from a school. The social service provision is slightly worse than for health centres, as 31.4% of the metropolitan population is between 5 and 10 minutes away from the nearest health centre, whereas 24.8% of the population has a similar journey time to reach the closest social services (Graph 2).

In view of these accessibility indicators, it is evident that public and subsidised schools are services with good accessibility for the highest percentages of the metropolitan population, whereas the situation of the health care and social services presents more problems, in principle. Nevertheless, generally speaking the provision of services in the MAV can be considered good, as most of the population residing in the area can reach a public service by public transport in under 30 minutes. Evidently, any improvements in accessibility should address transport to hospitals and bringing basic services closer to the population, in other words, expanding them in the most isolated and disadvantaged zones, as will be discussed here below.

	Population under 5 minutes away	Population 5-10 minutes away	Population 10-20 minutes away	Population 20-30 minutes away	Population 30-45 minutes away	Population 45-60 minutes away	Population over 60 minutes away
Hospitals	1.02	3.94	19.81	42.35	62.18	61.41	69.80
Health Centres	13.05	24.88	29.52	14.78	11.95	14.91	9.65
Social services	12.69	19.71	33.15	24.27	11.07	7.43	7.69
Primary schools	45.13	21.31	5.96	7.30	6.07	7.15	5.20
Secondary schools	28.11	30.17	11.56	11.31	8.74	9.09	7.66
Total	100	100	100	100	100	100	100

Table 3. MAV population by journey time to services (vertical percentages)
(Source: Own elaboration)

	Population under 5 minutes away	Population 5-10 minutes away	Population 10-20 minutes away	Population 20-30 minutes away	Population 30-45 minutes away	Population 45-60 minutes away	Population over 60 minutes away	Total
Hospitals	1.30	4.97	23.75	20.95	23.98	10.94	14.12	100
Health Centres	16.68	31.41	35.38	7.31	4.61	2.65	1.95	100
Social services	16.22	24.88	39.74	12.00	4.27	1.32	1.55	100
Primary schools	57.67	26.90	7.15	3.61	2.34	1.27	1.05	100
Secondary schools	35.93	38.08	13.86	5.60	3.37	1.62	1.55	100

Table 4. MAV population by journey time to services (horizontal percentages)
(Source: Own elaboration)

Graph 1. Population by journey time from hospitals

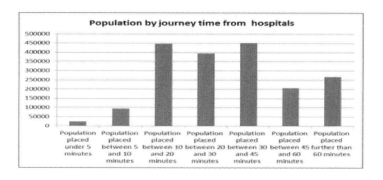

Graph 2. Population by journey time from health centres

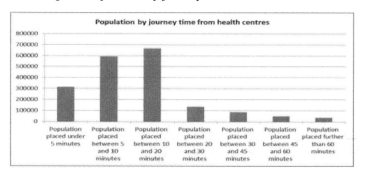

Graph 3. Population by journey time from primary schools

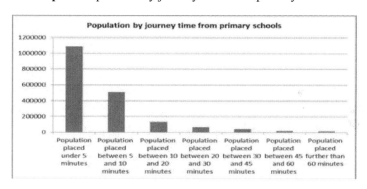

Graph 4. Population by journey time from secondary schools

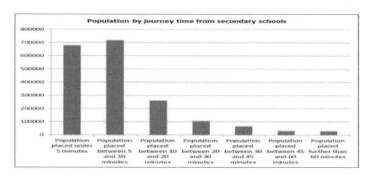

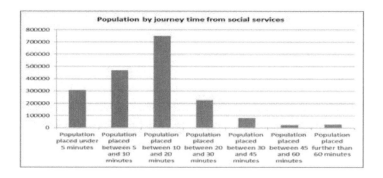

Graph 5. Population by journey time from social services

6.2. Areas of Influence of the Public Services

In terms of territorial variations in the accessibility of the public services studied, the five maps based on the Spatial Separation Index (ISE) calculations (Map 4) make it possible to draw two fundamental conclusions. Firstly, in general there is a centre-periphery type spatial pattern in which the accessibility levels are highest at the centre of the metropolitan area and decrease in concentric rings towards the periphery. This arrangement is particularly clear-cut in the case of the ISE values for health and education services, although the latter seem to combine the concentric rings with radial structures that coincide with the main routes (public and private) towards the north-west and west, matching the MAV's main areas of urban growth. The concentration of service provision in the larger population centres is explained not only by the generic location of the facilities (greater density in the central areas) but also by the greater quality (frequency and density of routes) of their public transport, particularly in the city of Valencia.

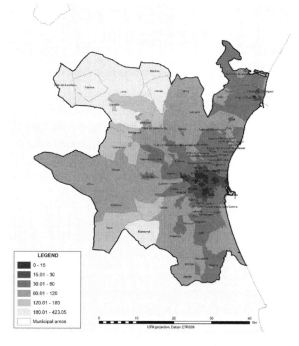

Map 4. Time in minutes (ISE) to the nearest hospital

Secondly, the metropolitan space is not homogenous, particularly on its periphery. It is not homogeneous either from the urban development point of view or from that of accessibility.

Accessibility does not diminish in the same way or at the same rate in every direction. This can be seen on some of the maps of ISE values, particularly the one for social services. Two axes of high accessibility are clearly visible there: a north-west axis and a north axis, which converge on the urban nucleus of the city of Valencia. This part of the MAV contains the spaces with the greatest access to public services and therefore, presumably, the greatest potential wellbeing; in short, better conditions for sustainability and equity. These spaces are favoured by both a denser provision of public facilities and public transport of greater quality.

6.3. Spaces of Inequality: Privileged Zones and Problem Territories

The method applied makes it possible to detect the zones with the best and worst access to public services within the area studied. In general, as was to be expected, the centres of the different urban areas, particularly the city of Valencia, showed higher accessibility values for all types of services. The first element that needs to be taken into account is access to the public transport network itself, as the zones with a worse transport service will also present a worse accessibility indicator for the other types of service (ISE).

In peripheral census tracts or neighbourhoods or in ones with a scattered population the opposite is the case: poor accessibility, a general lack of public service provision nearby and on occasion even a lack of public transport. The less advantaged zones from the point of view of accessibility respond to two socio-territorial models. One is census tracts in high-income districts, in other words, suburbs with a structure of family houses at a distance from urban centres where the predominant means of transport is private vehicles. The other is census tracts in run-down, low-income neighbourhoods. Some are on the periphery of the metropolitan area but others are close to the centre.

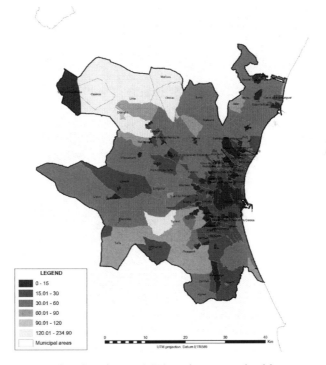

Map 5. Time in minutes (ISE) to the nearest health centre

In the first of these types, access by public transport is not a priority as owing to the high income of the inhabitants it is assumed that private vehicles will be used most, although that is not sufficient reason to justify a total lack of access to the public transport network. About 100,000 inhabitants of the MAV are in this situation. Reasons of sustainability, energy efficiency and safety clearly justify an adequate service for these zones, all the more so since their residents are known to be more mobile than average. The reason is that as these districts are mainly residential, their inhabitants' places of work are normally at a distance that requires the use of some form of transport. Another reason for their greater tendency to daily mobility is that both the services studied here and those related with shopping and leisure are also located some way away.

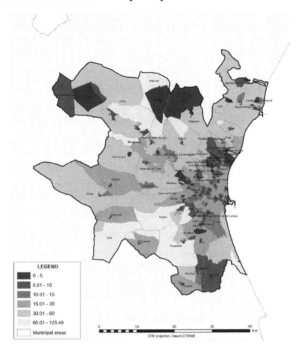

Map 6. Time in minutes (ISE) to the nearest basic social services centre

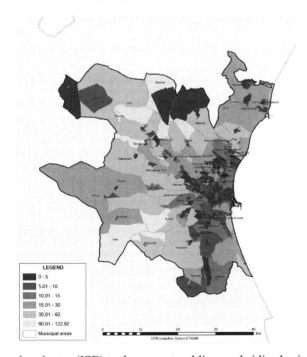

Map 7. Time in minutes (ISE) to the nearest public or subsidised primary school

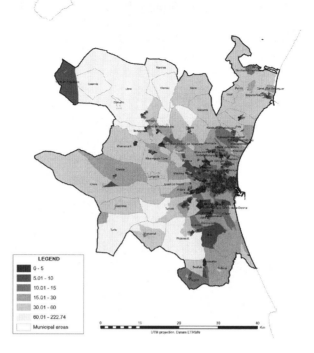

Map 8. Time in minutes (ISE) to the nearest public or subsidised secondary school

Census tracts without any type of public transport at a distance from the centre, in peripheral municipalities without access to the public transport network, account for a small proportion of the metropolitan area's population. The fact that most of the services are located in their population centre makes it unnecessary to travel elsewhere, except in some cases such as going to a hospital, which requires a journey as these are centrally located. In that event this 27,000-strong population needs to use private transport, which can be a problem for many, such as old people who do not own a vehicle.

Lastly, census tracts without public transport in urban areas are found in socially and territorially peripheral neighbourhoods. The population of the zones in this situation is slightly over 200,000, making this the most populous group of the three. Almost 20% of them live in the main city, Valencia, and over 30% in large towns or cities: Torrent, Sagunt, Mislata and Manises. As these are urban census tracts it is not impossible for their inhabitants to access the public transport network, although it is further away than in other neighbourhoods or tracts. Their isolation is considerable but not total (Map 9).

On combining the ISE results for the different public services with the existence or otherwise of a public transport network nearby, very different results appear for the different census tracts. Even within the same municipality the situation can be very heterogeneous. The accessibility of public services is better the closer the census tract is to a major urban centre and worse, generally, in the second metropolitan ring, which is more rural and further away from the city of Valencia and other cities and towns in the MAV.

However, some census tracts in the main cities and towns present a similar situation to that of the rural areas, despite their urban setting. They are usually but not always at a distance from the centre, have no public transport services or very unreliable ones and no provision of the type of public service in question. Using different criteria, it is possible to establish levels of inequality or scarcity for the different types of service. Schools are the service with the best distribution over the area, so very few of the population are affected (Table 6), while poor access to health services, particularly hospitals, causes problems for a higher percentage of the population. In the case of hospitals, 10% of the inhabitants of the MAV have to travel for over 30 minutes by public transport and for 3% the

journey takes over an hour.

Map 9. Census tracts with no public transport stop

	Hospitals	Health Centres	Social services	Primary schools	Secondary schools
Over 30 minutes away	198,830	29,017	22,882	17,407	27,228
Over 60 minutes away	59,904	13,192	8,386	9,117	13,192

Table 5. Number of people in the MAV with poor public transport access to public services

Locating the population that is affected by a greater scarcity of both public transport and public services is essential in order to identify where to invest in improving their provision and building new facilities. Although the overall figures for the MAV are not unfavourable, attention needs to be paid to zones which are no less important than others despite only accounting for a small proportion of the total population (Table 6). It is possible to identify the census tracts with the worst results for all the services studied.

	Number of census tracts	Population	% of total MAV population
Over 60 minutes from ALL the public services	10	16,934	0.8
Over 30 minutes from ALL the public services	40	69,206	3.6
Under 30 minutes from ALL the public services	588	909,759	48.2
Under 15 minutes from ALL the public services	137	180,086	9.5

Table 6. Impact of the accessibility by public transport of all the public services

It is useful to identify the worst-served census tracts, but also to find out which districts are best-positioned for access to the public service network, in other words, the location of equity. The worst-situated zones are, as mentioned, those that do not have a nearby public transport stop either, but they are not the only ones. Some census tracts are so large that the public transport network access point is at some distance from where part of the population lives.

In short, based on the accessibility indicator employed, equity is fairly adequate in the MAV. However, the best-served districts are those of the city of Valencia. Most census tracts that are less than 15 minutes away from all the services are in the central city or in municipalities that form part of its conurbation, although they only account for 9.5% of the population of the metropolitan area. The data are even more positive for journey times of 30 minutes at most. Over half the population is less than half an hour away from all the public services.

The worst situation is found in the municipalities on the periphery. The ten census tracts that are over an hour away from all the services belong to municipalities that are not part of the Valencia conurbation, are all at a distance from the urban centre of their own municipality and are of the dispersed habitat type, at times combined with weekend homes. The population in this situation comprises under 1% of the total inhabitants of the MAV. It is no less important for that, and at least as regards basic services (primary schools and health centres) its situation should be improved considerably. Quantitatively, the population that lives over 30 minutes from all services is larger. It makes up less than 4% of the total and also corresponds to dispersed habitat areas. The city of Valencia barely figures in this group, but the largest municipalities in the first metropolitan ring do. These are census tracts with low population densities at a distance from the town centre, where private transport predominates, as does the use of private services, in consonance with their income levels.

The differences by type of service are also interesting and noteworthy. The situations mentioned above are the extremes, the cases of best and worst accessibility of all the services. However, situations where there is adequate access to one service but not to the others abound and the combinations are very varied. Half the population is to be found in this somewhat unbalanced position, particularly as regards basic health care and social services.

7. CONCLUSION: SOCIAL EQUITY WITHIN THE TERRITORY. A MEASUREMENT OF INEQUALITY

The method explained and applied in this article shows considerable potential. The most laborious aspect of it is undoubtedly drawing up a GIS that includes the necessary information on the different territorial entities and elements. The basis is the location of the services and the structure of the transport network. This is completed with the most detailed demographic, social and economic information possible regarding the territory. Socio-demographic information is available in Spain at census tract level, considered the smallest territorial unit that does not breach statistical confidentiality. If this information were available at street block or, even better, housing level, it would bring a substantial qualitative improvement in the results of the model.

At all events, measuring equity through access to the education, health and social services by public transport provides very reliable results even when the exact location of the demand (the population) is not available. This is compensated for to a certain extent by the correct location of the centre offering the service. The indicator of real-time accessibility is extraordinary suitable for drawing closer to the real situation in a complex territory such as a metropolitan area, making it possible to arrive at conclusions that could not have been reached with a less precise method.

Once the various ISE limits have been established, the different zones and municipalities can be classified according to their greater or lesser equity. Setting these limits is an important aspect for public policy-making at municipal level or, more appropriately, at metropolitan level. In the case under study, the authorities can take action in two ways: they can provide or relocate the centres that offer these services and they can act on the public transport network. Optimising the former and expanding the latter so that it reaches most of the territory would bring a considerable improvement in the area's equity. In the case of the MAV the provision of public services is broad and varied, with only a few exceptions, and extending the public transport network would improve the journey times.

Evidently, the structure of the public transport network determines the results regarding equity, but that is the whole point. A more detailed analysis of the demand would highlight the worst-affected social groups, generally children, young people and old people, the main users not only of public transport but also of the most important services. Consequently, a study of the socio-economic and demographic characteristics of the neighbourhoods with the worst levels of equity would give a deeper insight into the real impact of the worst accessibility on particular population groups.

In short, this method holds out many and varied possibilities for the future. One of the most interesting vistas it opens up is the ability to run simulations to measure the consequences of new locations or closures, which is highly relevant in the current economic climate. The prospect of closing some centres may not necessarily be negative if the service is not reduced and is located efficiently and, above all, if territorial equity is borne in mind: it may even be improved.

REFERENCES

Albertos, J.M., Noguera, J., Pitarch, M.D., Salom. J. 2007. "La movilidad obligada en la Comunidad Valenciana entre 1991 y 2001: cambio territorial y nuevos procesos". Cuadernos de Geografía, 81-82, 93-118.

Bhat, C., Handy, S., Kockelman, K., Mahmassani, H., Chen, Q. and Weston, L. 2000: Development of an Urban Accessibility Index: Literature Review. Centre of Transportation Research, The University of Texas at Austin.

Bueno Abad, J.R., Perez Cosin, J.V. et al. 2008. Observaquart. Informe sobre la percepción social de los servicios socioculturales. Valencia: PUV.

Calvo, J.L. et al. 2001. "Análisis, diagnóstico y ordenación de equipamientos mediante formulaciones cartografiables: valoración de la accesibilidad y requerimientos de la asistencia hospitalaria en la CCAA de la Rioja mediante la técnica de potenciales". Berceo 141, 247-268.

Castells, M. 2001. "La crisis de la sociedad de la red global: 2001 y después". Anuario internacional CIDOB 1, 15-19.

Cebollada, À. , Avellaneda, P.G. 2008. "Equidad social en movilidad: reflexiones en torno a los casos de Barcelona y Lima". Diez años de cambios en el Mundo, en la Geografía y en las Ciencias Sociales, 1999-2008. Actas del X Coloquio Internacional de Geocrítica. Universidad de Barcelona, 26-30 de mayo de 2008.

Cervero, R. , Kockelman, K. 1997. "Travel Demand and the 3Ds: Density, Diversity and Design". Transportation Research D, 2.3, 199-219.

Corral Sáez. C. 1994. "El centro de la ciudad en las periferias". Ciudad y Territorio. Estudios Territoriales. II (100-01), 421-432.

Crooks, V.A. , Andrews, G.J. 2009. Primary health care: People, practice, place. Ashgate Publ. Ltd.

Davouidi, S. et al. 2009. "El desarrollo territorial: entre la perspectiva ambiental, la cohesión social y el crecimiento económico". In Feria. J.M. et al (eds.) Territorios. Sociedades y Políticas. Sevilla: Universidad Pablo de Olavide and AGE.

Domanski, R. 1979. „Accessibility, efficiency and spatial organization". In Environment and Planning A, 11(10) 1189-1206.

European Commission 2005. Measuring progress towards a more sustainable Europe. Sustainable development indicators for the European Union. Brussels, http://epp.eurostat.ec.europa.eu/cache/ITY_OFFPUB/KS-68-05-551/EN/KS-68-05-551-EN.PDF (viewed 10 August 2011).

European Commission 2009. Sustainable development in the European Union. Brussels: Eurostat, European Commission, http://epp.eurostat.ec.europa.eu/cache/ITY_OFFPUB/KS-78-09-865/EN/KS-78-09-865-EN.PDF (viewed 10 August 2011).

Feria, J.Mª 2004. "Problemas de definición de las áreas metropolitanas españolas". Boletín de la Asociación de Geógrafos Españoles n.38, 85-99.

Foster-Bey, J. 2002. Sprawl, smart growth and economic Opportunity. Washington: The Urban Institute.

Garcia Herrero, G. (Coord.) 2004. "Ciudades socialmente sostenibles". Grupo trabajo 19, VII

Congreso de CONAMA, Madrid, http://www.conama.org/documentos/GT19.pdf. (viewed 02 May 2011).

García, J.C. 2010. "Urban sprawl and travel to work: the case of the metropolitan area of Madrid". Journal of Transport Geography, 18, 197-213.

Garner, B.J. 1971. "Modelos de Geografía Urbana y de localización de asentamientos". In Chorley, R.J. and Haggett, E. (eds.) La Geografía y los modelos socioeconómicos. Madrid: IEAL.

Garrocho, C. , Campos, J. 2006. "Un indicador de accesibilidad a unidades de servicios clave para ciudades mexicanas: fundamentos, diseño y aplicación". Economía, Sociedad y Territorio, VI (22).

Gordon, P. , Richardson. W.H. 1997. "Are compact cities a desirable planning goal?". Journal of the American Planning Association, 63 (1): 95-106.

Harvey, D. 1973. Social justice and the city. London: McMillan.

Hine, J. 2003. "Social exclusion and transport systems", Transport Policy, 10: 263.

Holmes, J.H. 1985. "Policy issues concerning rural settlement in Australia's Pastoral zone". Australian Geographical Studies, 23-3, 3-27.

Mallarach, J. ,VIlagrasa, J. 2002. "Los procesos de desconcentración urbana de las ciudades españolas". Ería. Revista de Geografía n.57, 57-70.

Massam, B.H. 1980. The right place. New York: John & Son Inc.

May, J. , Thrift. N.J. (eds.) 2001. Timespaces. Geographies of temporality. London: Routledge.

Mérenne-Shoymaket, B. 1996. La localisation des services. Coll. Géographique d'aujourd'hui. Paris: Ed. Nathan Université.

Miralles, C. 2002. Transporte y ciudad. El binomio imperfecto. Barcelona: Ariel.

Miralles, C. 2011. "Dinámicas metropolitanes y tiempos de la movilidad. La región metropolitana de Barcelona, como ejemplo". Anales de Geografía. 31,(1) : 125-145.

Montero-Serrano, J., Bosque, J. , Romero-Calcerrada, R. 2008. "Cuantificación y cartografía de la sostenibilidad social a partir de tipologías urbanísticas". In Hernández, L. and Parreño, J.M. (eds.). Tecnologías de la Información geográfica para el desarrollo territorial, 76-91. Las Palmas de Gran Canaria: Servicio de Publicaciones y Difusión científica de la ULPGC.

Moreno, A. 1992. "Modelos para el estudio y previsión de la demanda de servicios colectivos". Actas del V Coloquio de Geografía Cuantitativa, 501-514. Universidad de Zaragoza.

Moreno, A. 1987. "Planificación especial de equipamientos públicos: el diagnóstico". X Congreso Nacional de Geografía, 2, 357-366. Departamento de Geografía, Universidad de Zaragoza.

Mückenberg, H. 2009. Familia, política del temps i desenvolupament urbà. L'Exemple de Bremen. Barcelona: IERMB.

Nel.Lo, O. 2001. Ciutat de ciutats. Reflexions sobre el procés d'urbanització a Catalunya. Barcelona: Empúries [Translated into Spanish: Cataluña, ciudad de ciudades. Lleida: Milenio, 2002.

Rosero-Bixby, L. 2004. "Spatial Access to health care in Costa Rica and its equity: a GIS-based study". Social Science and Medicine, 58 (7) : 1271-1284.

Rode, P., Burdett, R., Brown, R., Ramos, F., Kitazawa, K., Paccoud, A. , Tesfay, N. 2009: Cities and Social Equity: inequality, territory and urban form. London: Urban Age Programme, London School of Economics and Political Science. Available from http://lsecities.net/publications/research-reports/cities-and-social-equity/.

Schuurman, N. et al. 2010. "Measuring potential spatial access to primary health case physicians using a modified gravity model". Canadian Geographer, 54: 29-45.

Schwanen, T., Dieleman, F.M. , DIJST, M. 2001. "Travel behavior in Dutch monocentric and policentric urban systems". Journal of Transport Geography, 9:173-186.

Tsour, K.W., Hung, Y.T. , Chang, Y.L. 2005. "An accessibility-based integrated measure of relative spatial equity in urban public facilities". Cities, 22 (6): 424-435.

"DTH-1.0": TOWARDS AN ARTIFICIAL INTELLIGENCE DECISION SUPPORT SYSTEM FOR GEOGRAPHICAL ANALYSIS OF HEALTH DATA

Dimitris Kavroudakis
University of the Aegean, Geography Department, Mytilene, Lesvos, Greece, 81100,
www.aegean.gr , dimitrisk@geo.aegean.gr

Phaedon C. Kyriakidis
University of the Aegean, Geography Department, Mytilene, Lesvos, Greece, 81100,
phkyriakidis@geo.aegean.gr , http://www.geo.aegean.gr/greek/CVs/kyriakidis.htm

Abstract

Organizations such as hospitals, hold a great number of datasets which consist of many individual-based records. Artificial Intelligence methodologies incorporate approaches for knowledge retrieval and pattern discovery, which have been proven to be useful for data analysis in various disciplines. Decision trees methods belong to knowledge discovery methodologies and use advanced algorithms for the extraction of data patterns. This work describes the development of an autonomous Decision Support System ("Dth 1.0") for the real-time analysis of health data with the use of Decision Trees. The proposed system uses a patient dataset and prepares reports about the importance of the characteristics that determine the number of patients of a specific disease. Also, we describe the basic concept of Decision Trees, describes the design of a tree-based system and use a virtual database to illustrate the classification of patients in a hypothetical intra-hospital case study.

Keywords: *decision making, geographical analysis, artificial intelligence, data mining, health geography, decision Trees*

1. INTRODUCTION

The complexity of modern data analysis is constantly increasing as the number of variables involved increases. Modern scientific problems require even greater computational power to analyse available datasets to produce outputs for analysis and informed decision making. The more the variables and the characteristics of a problem, the greater the complexity between a scientific problem and its characteristics. A modern group of methodologies is Artificial Intelligence (AI) which was introduced during the last decades in computational sciences. This group of methods include Machine Learning (ML) which is a category of methods for the extraction of knowledge from data to "train" a system which will later accumulate knowledge for analysis and prediction. The adaptation of AI methods to problem solving and data analysis is important to modern scientists as complexity of scientific problems increases. Moreover, geographical applications require advanced tools for spatial analysis. The variety, type and computational intensity involved in spatial data analysis, make AI methods valuable to the modern scientific arsenal of geographical analysis methods. Data mining is a methodological group of AI which extracts information from data. Often the type, extent and complexity of datasets hides the underlying information and trends which are crucial for scientific analysis. Those trends and associations can be extracted with data mining operations. "Decision trees" is a common modern data mining methodology dealing with training and prediction. This approach predicts the value of a variable by knowing other available attributes and can be applied to variables of categorical nature. For example when individual based data are available, then by training a Decision Tree model, a prediction model can be build to predict future occupancies based on statistical probabilities of other individual information. A good example of challenges and capabilities of the scientific field of data mining is the work of Witten et. al. (2011) which analyses ML and data mining and provides practical recommendations. Health geography is a field of geography focusing on the spatial characteristics of health related problems. Some of the topics of health geography can be approached by the use of AI methods. For example Decision Trees can be implemented towards the development of a health decision support system to understand and illustrate possible development of a disease by a patient via the use of patient's health record database. In other words a Decision Tree can be implemented for understanding the levels of a variable (development of disease) by parsing other data on the system. The underlying mechanism associates habits with a disease and offers a statistical model for the analysis of the development of the disease. This work illustrates the use of Decision Tree models for the understanding of health related datasets. After the presentation of the theory of Decision Trees, a model is constructed. Following the generation of a random health dataset ("arth2000") the model is trained to predict the development of a disease in a sample of 1000 patients. Finally we discuss the potential use and extensions of the proposed decision support system.

2. DECISION TREE CLASSIFICATION

Classification is a methodology of Machine Learning, useful for conceptualizing the classification of data groups. It assigns class labels to cases based on models linking known class labels with attribute levels. Some of the most common data classification techniques are: neural networks (Haykin 1994, Hagan et al. 1996), Bayesian networks (Rubin 1981, Jensen 1996, Jensen et al. 2007, Darwiche et al. 2009) and Support Vector Machines (Cristianini et al. 2000, Hearst et al. 2002, Moore 2007, Steinwart et al. 2008). Those approaches have advantages and disadvantages and are suitable for different type of analysis. Decision tree methodology is mostly suitable for classifying datasets of nominal variables and exploration of relationships between standardized variations of data attributes. Some of the advantages include the relatively fast learning algorithms that can be used such as ID3 (Cheng et al. 1988) and C4.5 (Quinlan 1993) and the robustness of the methodology to data noise such as missing values and attribute noise. Some of the disadvantages of Decision Trees include difficulty to represent the parity of values in a relationship and proportional complexity of the output diagram which sometimes can be misleading if not followed by expert analysis. This is when data is complex and the output mathematical graph has a substantial amount of nodes and leafs, human eye can be misled. This will not happen if the tree information graph is followed by a detailed explanation of each part of the tree and the meaning of the results. Decision tree approaches accept a dataset of nominal data and produces a dendrogramatic representation (Phipps 1971) of the data variables according to their levels. The steps of the algorithm used for such a classification are the following:

1. A is the best decision attribute for the next node
2. Assign A as decision attribute for the node
3. For each value of A, create a new descendant of the node
4. Sort data by leaf nodes
5. Iterate over leaf nodes, until data are classified

The result of this algorithm produces a number of nodes and leafs illustrating the number of potential decisions from the attributes of the data. This categorization of potential decisions groups all possibilities and counts the most prominent ones. The modelling process splits data into two subsets: one for learning and the other for prediction. The first subset of the data is used as a trainee for the model and is used for training the model to understand the relationship between attributes such as for example: medical illness and age group of the patient. The second data subset is used for evaluating this knowledge. The error calculation is based on the number of successful predictions of the model over the total number of cases. The ratio between training and evaluation data, depends on the total number of cases and on the type and extend of the individual-based data attributes. The error quantification depends on the number of cases, the number of variables and the number of levels in the ordinal scale, discretizing the range of variability of continuous explanatory variables.

Decision trees can process data involved in various disciplines and develop a knowledge discovery tool-set to predict levels of ordinal dependent variables. In the broader scientific field of health sciences there have been some interesting attempts to use Decision Tree methodologies. The work by Andreescu et. al. (2008) illustrates the use of Decision Trees in the prediction of patients respond to treatment of late-life depression. With a number of 461 records, the authors developed a hierarchy of predictors with Decision Trees. Additionally, the work of Mann et. al. (2008) focus on determining the risk for a suicide attempt in psychiatric patients from the analysis of multiple individual-based risk factors. Decision tree method has been used in a health dataset of 408 patients with mood schizophrenia or personality disorders to distinguish possible attempters. Another interesting use of Decision Trees in health sciences is the work by Zhang et. al. (2010) which is an attempt to demonstrate the effectiveness of one treatment against another with respect to pregnancy in polycystic ovary syndrome (PPCOS). That work used a dataset of 445 women who ovulated in response to treatments among a dataset of 626 participants. Decision tree approach was used to reflect treatment results between types of the syndrome. Furthermore, Koko et. al. (1998) described the evaluation of various Decision Tree methods on problems of orthopaedic fracture data and concluded that there are some limitations on the accuracy of the model and the sensitivity of the Decision Tree size. Tsien et. al. (1998) in their research about classification trees for the diagnosis of myocardial infraction, concluded that ML methods such as Decision Trees can be used in medicine for supporting early diagnostic decisions. Jones et. al. (2001), illustrate the use of Decision Trees in the identification of signals of possible drug reactions and concluded that data mining methods, such as Decision Trees can be a promising tool for identifying new patterns in medical datasets. Dantchev et. al. (1996) argue that Decision Trees are still in experimental stages and remain difficult to apply to clinical practice in psychiatry. In the same report they argue that those tools allow researchers to see epidemiological data from a more generalized perspective and focus on new priorities. Letourneau et. al. (1998) focused on decision making techniques for chronic wound care and concluded that Decision Trees can help decision making by guiding trained personnel through assessment and treatment options. Another notable example of using Decision Trees in health-related sciences can be seen in the work of Alemi and Gustason (2006), who describes some analytical tools for informative decision making. Decision Trees have been used in the broader health-related sciences and appears to be a methodological approach for informed decision making.

3. USE OF DECISION TREES METHOD

The following part describes the use of Decision Trees in the proposed health decision support system. Decision tools have been use the later decades for supporting decision making and help decision makers to get informed decisions. This part will illustrate the use of a decision making tool which included a Decision Tree approach. Initially an artificial health-database ("arth2000") is used for the training and validation of the method. The dataset consists of five independent variables (city, age, sex, activity, the milk) and one dependent variable (hypothetical disease AD). The cases of the dataset represent visits of individuals to a health care facility (hospital). For the purpose of this work, it is assumed that during each visit the medical personnel examined the individuals and recorded information about those specific variables. The total cases in the dataset is n=2000 (1000 training and 1000 evaluating cases) as can be seen in table 1. Table 2 shows the probabilities for each variable's level, used for the generation of the "arth2000" dataset. Those probabilities are also used for the evaluation of the accuracy of the classification. For the purpose of this work we assume that independent variables of the "arth2000" database, determine the values of the dependent variable and via the Decision Tree methodology we illustrate the type and extend of this influence. Subsequently, the model will be able to predict the probability of existence of the AD disease by processing the levels of the five dependent variables of an individual. The process of prediction can be complex and may also depend on the type and quality of the available dataset. The variables of the data include: the city variable which indicates the area of the patients permanent residence (5 levels), the age variable is a categorization of the age group of the patient (4 levels). Activity variable indicate the type of patient's active or passive type of living (2 levels) and milk consumption variable indicates the daily amount of milk consumption from the patient (3 levels). We prepare a model which uses information gain as a quality measure to populate a dendrogram. Information gain is represented as the entropy value of the data passed from the model. During the data training process, new data values add up to the overall information scheme. This information-gain measure is related to the number of previous occurrences of a particular combination of data values, in the dataset. For example, if the model is trained with a list of 10 individuals consisting of 9 males and one female, the information gain increases by one when the model is processing the 10th individual because up until the 9th individual the model only knew about the existence of just one sex-level (male). After handling the new sex-level (female) the model creates a new category of individuals and assigns all new females to that.

City of residence	Age category	Sex	Daily activity	Milk consumption	Suffer from AB disease
Athens	mature	female	high	high	Yes
Mytilini	old	female	average	average	Yes
Mytilini	mature	male	low	low	No

Table 1. Overview of the „arth2000" dataset

city		age		activity		milk		AD disease	
Athens	0.1	middle	0.2	active	0.5	average	0.3	No	0.4
Chios	0.2	old	0.4	passive	0.4	high	0.2	Yes	0.6
Crete	0.2	young	0.1			low	0.5		
Mytilini	0.3	mature	0.3						
Rodos	0.2								

Table 2. Probabilities of levels for the variables of the „arth2000" dataset

Additionally the model uses no pruning mechanism as this would limit the extent of the tree and because the number of cases in the "arth2000" dataset is limited. The minimum records per node is set to 5 which means that the information gain weight should reach 5 individual records before creating a new leaf. This is an empirical value and varies depending on the type of data or analysis and the required results complexity. Figure 1 depicts the overall theoretical structure of the developed model. Initially, health data are inserted via a csv reader. Those data could be in principle fetched from the examination rooms of a hospital and following the model processing procedure, can be presented to decision makers in real time. The proposed procedure then splits the database in two parts for cross validation. In other words, the model will learn from the first part of the data and then evaluate the quality of the knowledge upon the second half of the data. The process of learning is held in module number 3 and the linkage of the dataset is taking place in module number 4. The linkage evaluates the leaning ability of the model with the remaining data.

Decision Trees approach, helps solve a problem, which in this case-study is the understanding of possible future existence of disease AB to a number of patients by knowing a limited number of information about each patient. This approach is used to represent the various decision points along the examination of a potential patient. As can be seen in table 3, (Rule 3), if the patient is from Chios city, it has an increased probability to suffer from AB disease. In that case, the examination personnel, then needs to ask the patient about his/her age as in Rule 6, patients of mature, old and young age, have an additional increased probability to suffer from disease AB. The resulted Decision Tree, offers a list of characteristics which have increased probability over the possible existence of AB disease, according to the "arth2000" dataset. It describes the logical steps required for determining whether an individual has increased probability to suffer from disease AB (dependent variable) by knowing the value of a number of other variables (independent variables).

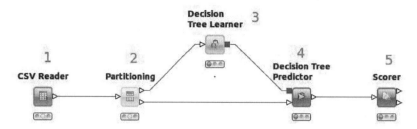

Figure 1. Work-flow process diagram for data mining of the arth2000 dataset

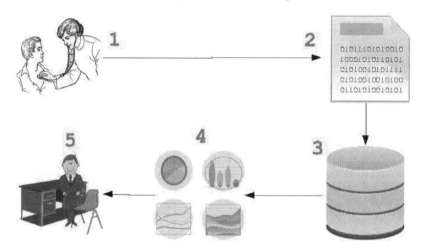

Figure 2. The abstract design of the proposed decision system

Rule				Appearance of disease AB	cases	percentage	categories
1				No	450/1000	45%	
	2			No	348/791	43%	milk=high,low
		4		No	239/569	42%	city=Athens,Crete, Mytilini, Rodos milk=high,low
		5		No	109/222	49%	city=Athens,Crete, Mytilini, Rodos milk=average
			10	No	51/120	42%	city=Crete, Rodos milk=average
			11	Yes	44/102	43%	milk=average city=Athens, Mytilini
	3			No	102/209	48%	city=Chios
		6		No	83/180	46%	city=Chios age=mature, old, young
		7		Yes	10/29	34%	city=Chios age=middle

Table 3. The categorization statements for the "arth2000" dataset

The Decision Tree methodology consists of a root node split by a single variable into partitions. In turn those partitions become nodes to be split further. This divide-and-conquer approach continues until no further splitting would improve the performance of the model. The performance is the ability of the model to understand the possible categorization of a case, based on its attributes. This ability of the model, increases as the statements incorporate additional knowledge about the training dataset. In other words, the more the information gain from a categorization, the higher the ability of the model to categorize cases with less information. The categorization statements of the produced Decision Tree are depicted in table 3. For each statement (row) the first column is the id of the statement. The disease AB column indicates that this statement can categorize cases that may suffer from disease AB. The "cover" column shows the number of cases that have been categorized with this rule. The percentage column shows the percentage of the categorized cases with respect to the total number of cases. Finally, the "categories" column shows the information gain from each rule. For example rule 11 indicates that cases from Athens or Mytilene with average consumption of milk, have 43% probability to suffer from disease AB. The combination of those rules can categorize this particular "arth2000" dataset with 100% accuracy. In order for the model to judge how good a potential split (node-leaf) is, the information gain rule is used, which creates a new split at the attribute with the highest information gain. This approach creates new splits only when they will create concrete partitions of the dataset. The split function strategy for this model is the "entropy reduction strategy". The greater the information from a categorization, the greater the knowledge of the model for future categorizations. Finally the minimum number of observations in a node before attempting a split (splitting factor) is 100 cases. This reduces the tree complexity and produces a more readable representation. The selection of the splitting factor depends on the number of total cases and the amount of detail required in the results. The more complex the results, the more the leafs of the Decision Tree.

The very same categorization process can be used with any other categorical dataset to prepare categorization rules. This makes the approach generalizable and flexible. A factor that determines the accuracy of the results is the number of levels for each variable. Binomial variables can be categorized with less rules than variables with 4 or 5 levels. This is making sense if we consider that entropy increases for variables with larger number of levels. This forces the model to produce more rules to fully categorize a dataset. On the other hand, binomial variables can be easily categorized and require less rules. Finally, a fully randomized dataset, where there is no relationship between variables, may require a great number of rules to be fully categorized. The tendency of the dataset towards a random distribution is directly associated with the inability of the Decision Tree to categorize all cases with less rules.

The methodology of data categorization discussed, can be potentially useful to health facilities, such as hospitals for categorizing patient records and present statistics based on patient profile. For example a large hospital with a great number of daily visits, can produce a good amount of data related to patient characteristics and health problems which can be obtained from a patient upon his/her arrival and stored in an internal database. Then, a centralized computational decision support system can process this dataset and prepare correlations between patient's characteristics. A Decision Tree can be implemented in R statistical language, with the use of the "rpart" library (Therneau et al. 2005), to prepare a statistical description for such a database. The administration of a hospital may access the daily results of the process and use visualizations indicating trends and patters that are not initially visible and can offer a centralized and categorized view over the characteristics of each disease. Informed decision making applications such as the one described, can help towards early warning of disease spread or help hospital personnel towards systematic diagnosis of symptoms.

4. AUTOMATED HEALTH DECISION SYSTEM

The proposed system makes decisions based upon health related datasets and evaluates the possibilities of occurrence of a control variable. The application of the proposed system is the statistical analysis of health databases from non-expert uses such as managers and directors of hospitals. This analysis can also be helpful in early examination procedures or in decision making in epistemological analysis. Considering the lack of centralized statistical tool in Greek Health Sector, this work is innovative as it describes the use of such a tool and advocates the use of not only descriptive statistics but also the use of AI for advanced statistical analysis. Sometimes it is difficult to fully understand the big picture of a cause-effect relationship especially when it is hidden deep into a great amount of data. The proposed system is using data mining methods and databases to construct a "hospital-oriented" computer system for the preparation of on-the-fly patient statistics. As illustrated in figure 2 the flow of information starts from the examination room (point 1) where the basic description of the characteristics and the medical record are transformed into a digital record (point 2) and inserted into a database (point 3). This process can be facilitated by modern Computer-Tablet hardware with live Graphical User Interface that can either create a new medical record or update an existing one with new medical information. The database of the system can retrieve data with queries. Instead for the researchers/managers (point 5) to use traditional SQL queries to retrieve tabular results, they can use a series of predefined actions that will call a number of queries. The pre-defined actions will present the results in graphical form (point 4). One of those predefined actions may include the use of a Decision Tree to analyse the data and observe the relationship between variables.

This proposed system requires a number of "examination-room computers" which will be used for data entry during examination. The software required for those machines, should be just basic intra-net browsing and an html compatible client (web browser). The main database will be hosted on an in-house server which will be an average computer with an installation of a database (PostGreSQL, MySql). In this computer the freely available statistical environment R is required which will prepare statistical analysis, and graphs. Finally the managerial and administrative staff will need computers with basic intra-net browsing for selection of predefined actions and visualization of results. The preparation of predefined actions saves time and effort and provides a relatively error-free environment for the generation of statistical results. The following code snippet (figure 3) shows an example of a predefined action which can be used by a managerial staff to prepare a description of the age categories of the patients in the database. The following code snippet (figure 4) is written in R statistical programming language and its output is depicted in figure 5.

- Lines 1-3 import the required libraries and packages for the analysis.
- Lines 5 and 6 activate the database driver and connect to the database.
- Lines 8 and 11 discover the available tables in the database and list its fields.
- Lines 14 and 15 retrieve all available data from the database table
- Lines 17-20 convert the data according to age categories
- Lines 22-24 prepare the diagram variables
- Lines 26-28 generate the bar-plot diagram

The generated bar-plot show the absolute number of patients and the percentage for each age group. This predefined statistical analysis generates output that informs the user (managerial and admin staff) about the age groups of the patients that visit the hospital. Another predefined statistical "action" could be the profiling of the dataset with Decision Trees methodology. This methodological approach may serve as an information categorization process that generates a dendrogram of groups of cases. Figure 4 depicts the R code of such a process which may be used in the proposed computational system. After the installation and calibration of this autonomous health decision support system, the users can explore statistical information of the patient's dataset without any knowledge on statistics or artificial intelligence. This is an autonomous system as it uses only the predefined statistical actions and does not require any input by the user. This enables the end user to focus only on a number of important and error-free statistical analyses. As with other similar systems, the quality of the imported data will influence the quality of the exported statistics. The proposed system may serve as business intelligence platform for health facilities and will be a valuable asset to epidemiological agencies.

```
1  library("RSQLite")
2  library("DBI")
3  library("tools")
4
5  drv <- dbDriver("SQLite")
6  db <- dbConnect(drv, "data.sqlite")
7
8  print(dbListTables(db))
9  #arth2000
10
11 print(dbListFields(db, "arth2000"))
12 #"id" "activity" "milk" "age" "city" "arth"
13
14 ola <- dbSendQuery(db, "select * from arth2000")
15 d <- fetch(ola, n=2000)
16
17 yo <- length(subset(d, d$age=="young")[[1]])
18 ma <- length(subset(d, d$age=="mature")[[1]])
19 ol <- length(subset(d, d$age=="old")[[1]])
20 mi <- length(subset(d, d$age=="middle")[[1]])
21
22 counts <- c(yo,ma,ol,mi)
23 per <- (counts/sum(counts))*100
24 names <- c("young", "Mature", "Old", "Middle")
25
26 mp <- barplot(counts, names.arg=names)
27 text(mp, counts + 50, format(paste(per,"%")), xpd = TRUE)
28 text(mp, counts - 50, format(counts), xpd = TRUE, col = "blue")
```

Figure 3. Snippet of R code for the generation of bar-plot with the age structure of the patients. The generated plot is figure 3

```
1  df=createData(2000)
2  library("rpart")
3  tree <- rpart(arth ~ city + activity + age + milk, method="class",
4                data=df,control=rpart.control(minsplit=100,cp=0.001))
5  summary(tree)
6  plot(tree)
```

Figure 4. Snippet of R code for the generation of sample data and the subsequent construction of a Decision Tree from them

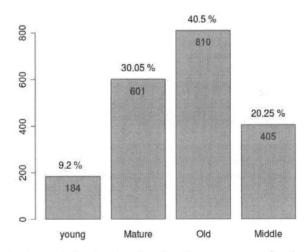

Figure 5. Result of a predefined action showing the age groups of patients in the dataset

5. GEOGRAPHICAL ANALYSIS OF HEALTH DATA

The discussed health dataset, include information about the place of residence of the patient. This is important as can provide spatial attributes to health records for further analysis. The geographical referenced health data can be presented in maps and analysed by area in order to provide a better insight on the spatial distribution of health events. Also proximity to health facilities can be examined and associated with the type of illness or the socio-economic characteristics of patients. This type of information could be used for geographical analysis of a disease outspread and map the areas which have significant amount of incidents. The understanding of spatial characteristics of a disease outspread can help decision makers to provide better health services and information to the public for the protection of public health. Geographical epidemiological studies aim to understand the spatial characteristics of health data and formulate hypotheses regarding the spatial causes and effects of a disease (Rezaeian et al. 2007). Some of the different branches of spatial epidemiology are disease mapping, cluster identification and spatial socio-economic analysis of a disease(Elliott et al. 2004). Understanding the greater spatial trends of a disease as well as mapping the spatial distribution of a disease from health records of a hospital, can be a challenging task especially due to privacy and ethical reasons. Nevertheless, if handle with case, geo- referenced health data can be very useful in early warning epidemiology systems and provide a different approach on understanding the causes and effects of a disease outspread in an area. Automated systems of identifying underlying data trends such as Decision Trees methods, provide the necessary layer of data mining, which extracts knowledge from individual records. The use of such an autonomous system can spare financial and other resources by avoiding the manual examination of data records and by quickly indicating the geographical characteristics of an event.

6. DISCUSSION AND CONCLUSIONS

The automated system discussed in this work may provide a basic mechanism for real-time analysis of health related data. The statistical analysis of databases may not be substitute for the actual medical diagnosis from specialized personnel. Nevertheless it is a first step towards the automation of health diagnosis and a valuable tool for decision makers and researchers to understand and identify underlying health data patterns. The "arth2000" dataset is an arbitrary dataset consisting of 2000 cases. No real data have been used in this stage of the research as there is lack of free health database for Greece. In a future state of this research, there is the possibility to use anonymized data from Greek Health Public System. One of the benefits of using Decision Tree methodology is that by scaling up the number of the cases and using a real medical dataset consisting of more than 10000 cases, the accuracy of the model will increase, and the model will be able to learn faster. Additionally, the geographical analysis of the model results can be facilitated by analysing the effects of disease AB by geographical area or city of residence which is the second determinant factor of disease AB. This may indicate that disease AB is more common in some geographical areas than others. This approach may be the basis for a more advanced spatial analysis and may lead to more sophisticated results. Such results may include the identification of the reasons and effects of a disease on other characteristics of daily life of individuals. Another interesting point is that this analysis can be automated. This is that an intelligent autonomus system can be constructed with very basic hardware such as a home-range personal computer which accepts data from a hospital and prepares instantly (in real time) statistical reports about the profile of the patients. Finally, this proposed system may be used for research and decision support processes and act as a data exploration tool for informed public policy decisions.

With the use of ML methods such as Decision Trees, one can better reveal underlying information which is hidden in data. Complex relationships that may exist in very large datasets are sometimes difficult to understand and may require a great number of computations and cross-tabulations. Analysis of rich and complex datasets can be valuable to the health sector as it may reveal underlying geographical patterns for diseases, symptoms and characteristics of patients. Finally, Decision Trees in general may act as a framework to consider the probability of events and pay-offs of decisions in various data analyses, not only in health-related sectors but also from in sectors such as geography marketing and logistics.

REFERENCES

Alemi, F. & Gustafson, D. H., 2006. Decision Analysis for Healthcare Managers, Health Administration Press.

Andreescu, C., Mulsant, B. H., Houck, P. R., Whyte, E. M., Mazumdar, S., Dombrovski, A. Y., et al., 2008. Empirically Derived Decision Trees for the Treatment of Late-Life Depression. Am J Psychiatry, 165(7), p. 855-862.

Cheng, J., Fayyad, U.M., Irani, K.B. & Qian, Z., 1988. Improved decision trees: A generalized version of ID3. In Proceedings of the Fifth International Conference on Machine Learning: June 12-14, 1988, University of Michigan, Ann Arbor. p. 100.

Cristianini, N. & Shawe-Taylor, J., 2000. Support Vector Machines, Cambridge University Press.

Dantchev, N., 1996. Decision trees in psychiatric therapy. Encephale-Revue De Psychiatrie Clinique Biologique Et Therapeutique, 22(3), p. 205-214.

Darwiche, A. & Corporation, E., 2009. Modeling and Reasoning with Bayesian Networks, Citeseer.

Elliott, P. & Wartenberg, D., 2004. Spatial epidemiology: current approaches and future challenges. Environmental health perspectives, 112(9), p. 998.

Hagan, M.T., Demuth, H.B. & Beale, M.H., 1996. Neural Network Design, PWS Boston, MA.

Haykin, S., 1994. Neural Networks: A Comprehensive Foundation, Prentice Hall PTR Upper Saddle River, NJ, USA.

Hearst, M.A., Dumais, S.T., Osman, E., Platt, J. & Scholkopf, B., 2002. Support vector machines. Intelligent Systems and their Applications, IEEE, 13(4), p. 18-28.

Jensen, F.V., 1996. An Introduction to Bayesian Networks, UCL press London.

Jensen, F.V. & Nielsen, T.D., 2007. Bayesian networks and decision graphs, Springer Verlag.

Jones, J.K., 2001. The role of data mining technology in the identification of signals of possible adverse drug reactions: value and limitations. Current Therapeutic Research, 62(9), p.664–672.

Kokol, P., Zorman, M., Stiglic, M. & Malcic, I., 1998. The limitations of decision trees and automatic learning in real world medical decision making. In Proc. 9th World Congr. Med. Inform. (MEDINFO-98). pp. 529-533.

Letourneau, S. & Jensen, L., 1998. Impact of a decision tree on chronic wound care. Journal of Wound, Ostomy, and Continence Nursing: Official Publication of The Wound, Ostomy and Continence Nurses Society / WOCN, 25(5), p. 240-247.

Mann, J. J., Ellis, S. P., Waternaux, C. M., XINHUA, L., Oquendo, M. A., Malone, K. M., et al., 2008. Classification trees distinguish suicide attempters in major psychiatric disorders: a model of clinical decision making. The Journal of clinical psychiatry, 69(1), p. 23-31.

Moore, A. W., 2007. Support Vector Machines. http://jmvidal. cse. sc. edu/csce883/svm14. pdf, 31, p.2001.

Phipps, J. B., 1971. Dendrogram topology. Systematic Biology, 20(3), p. 306.

Quinlan, J. R., 1993. C4.5: Programs For Machine Learning, Morgan Kaufmann.

Rezaeian, M., Dunn, G., Leger, S.S. & Appleby, L., 2007. Geographical epidemiology, spatial analysis and geographical information systems: a multidisciplinary glossary. Journal of Epidemiology and Community Health, 61(2), p.98–102. Available at: [Accessed May 28, 2013].

Rubin, D. B., 1981. The bayesian bootstrap. The Annals of Statistics, p.130–134.

Steinwart, I. & Christmann, A., 2008. Support Vector Machines, Springer Verlag.

Therneau, T. M., Atkinson, B. & Ripley, B., 2005. Rpart: recursive partitioning. R package version, 3, p.1-23.

Tsien, C.L., Fraser, H. S., Long, W. J. & Kennedy, R. L., 1998. Using classification tree and logistic regression methods to diagnose myocardial infarction. Studies in Health Technology and Informatics, 52 Pt 1, p.493-497.

Witten, I. H. & Frank, E., 2011. Data Mining: Practical Machine Learning Tools and Techniques 3rd ed., Morgan Kaufmann Pub.

Zhang, H., Legro, R. S., Zhang, J., Zhang, L., Chen, X., Huang, H., et al., 2010. Decision trees for identifying predictors of treatment effectiveness in clinical trials and its application to ovulation in a study of women with polycystic ovary syndrome. Human Reproduction, 25(10), p.2612-2621.

MAPPING GEOGRAPHICAL INEQUALITIES OF INFORMATION ACCESSIBILITY AND USAGE: THE CASE OF HUNGARY

Ákos JAKOBI
Eötvös Loánd University, Institute of Geography and Earth Sciences,
Department of Regional Science, Pázmány Péter sétány 1/C, H-1117, Budapest, Hungary
http://geosci.elte.hu/en_index.htm, jakobi@caesar.elte.hu

Abstract

In the era of information and communication technologies it is necessary to clarify what motives are in the background of disparities. The paper firstly sets up a theoretical framework that depending on the phase of innovation adaptation, main features of inequalities have different basic characteristics. By the multivariable analysis of information accessibility differences in the Hungarian microregions, basically the physical infrastructural constraints of the information society and economy are determined. Secondly, since an increasing number of people have become able to access the new information channels by today, the factor of accessibility could now be treated as a secondary problem. In contrast with accessibility differences the inequalities of usage come into the forefront of the analyses. Finally, the last section reveals that instead of usage volume differences the inequalities in the quality of information usage should be taken into consideration when dealing with the newest geographical inequalities of the information age.

Keywords: *information society, regional inequalities, ICT, GIS, Hungary, EUROGEO 2013*

1. INTRODUCTION – THE CHANGING NATURE OF ICT INEQUALITIES

Nowadays, alongside of traditional factors of inequalities some new ones seem to emerge, which have thorough effects also on geographical disparities. In connection with the currently very popular phrase of information society there are more and more practical experiences confirming that processes affecting regional differences are showing also new characteristics. For example by the appearance of the new innovations of information and communication technologies (ICTs) a transformation process has begun, which has changed our opinion on the advantages and disadvantages of geographical position, location, distance or other geographical factors. In information inequalities besides economic and social factors thus an increasing role is believed to be played by geography as well. Since recognising the growing importance of the notion of information society, modern geography has certainly the task to discover and evaluate the main characteristics of changes induced by the information age. Actually there is an increasing demand on clarifying what reasons are in the background of disparities. The explanation of the function of geography in information inequalities by the clarification of accessibility disparities and user differences could serve the better understanding of recent days' altering processes.

When speaking about linking traditions with future an obvious question arises within the circles of geographers: does future or recent geographical inequalities have the same basic characteristics, or new inequalities are different from traditional ones? In the context of information society this research question could be further specified: does geography of the information age really differ from geographies of previous times, or it has significant ties with traditional geographical concepts and research results? These questions practically reflect on the fact that we may look upon geographical problems a different way time by time. This time our eyes are focusing on geographical inequalities, while trying to determine what spatial characteristics are typical in information accessibility and usage differences.

Theoretically, among factors of geographical inequalities three basic types can be identified (Jakobi, 2004). The first covers traditional factors, which had same effects on geographical inequalities in the past as they will probably have in the future. The second group contains transformed factors, which were on the scene in the past, too, but have a different kind of influence on inequalities at present. Finally, the third type is about new factors, which either did not exist formerly or did not have any influence on the spatial inequalities, but exert a strong geographical impact nowadays. According to these groupings, at first sight modern age factors of information society should be considered as new inequality effects, since main innovations of ICT appeared only in the last couple of decades, or are appearing recently, and typically new type of geographical disparities, e.g. virtual space inequalities (Graham and Aurigi, 1997) or the digital divide (Mossberger et al. 2003) is related to it.

Information society as a distinctly new socioeconomic concept, motivation factor and value system burst upon the scene, though not without preliminaries, in the last few decades; it is a consequence of the social evolution caused by accelerated technological development. The commercial opening of the Internet at the beginning of the '90s brought really important changes into the previously almost closed world of the Web. The new innovation was spreading at an unprecedented rate, and has required a completely new way of thinking, which proved to be an extremely useful instrument and also created new opportunities. Although computers and mobile phones were in use much earlier and information and knowledge have always been important, the substantial change was due to all these elements being associated with the main production factors (besides labour force and capital), thus gaining much more importance than simple tools.

In truth the term "information society" has been used by researchers since the 1960s (Umesao, 1963; Porat, 1977) and has appeared from utopian to matter-of-fact scientific approaches in many contexts. It has had the highest occurrence among the keywords of publications in the last couple of decades (Masuda, 1988; Fichman, 1992; Castells, 1998; Trujillo, 2001; van Dijk, 2005). The research of ICT-based inequalities despite the novelty of this term is already not unknown in circles of international researchers. Basic works of Castells (1996, 1997, 1998), Norris (2001) or van Dijk (2005) formulated many concepts on inequalities of the information society. Also regional aspects of this topic became widely analysed (Goddard et al. 1985; Odendaal, 2003), however it has still a lot of questions to be answered, especially in relation with cyberspace inequalities and those effects on traditional geographical features. Joining to this, actually, it seems to be a re-emerging question, whether ICT-based inequalities are typically new ones, or they are just reshaping existing differences.

Digital divide or the digital gaps are the expressions of the researchers of information society on describing how specific the inequalities are in this environment. In the background of ICT-based differences there are (also) general social distinctions, namely income, education, gender or age differences of the population (Servon, 2002), which are basically traditional inequality factors even in the information age. We should note that digital divide cumulatively foster existing social inequalities, therefore in that sense the factors of information inequalities may be considered also as traditional, or at least transforming ones, although ICT still have in majority new type of differentiating effects. All in all, if we go into details, we might discover that there are traditional, altering and new inequality motives within the topic of information age disparities, too.

To prove this concept, first of all the changing nature of ICT-based inequalities should be understood. Models, which try to explain and quantify inequalities of information accessibility and usage, evaluate the factors of early and late phases of technological development differently. Professionals explain the altering role of the influencing effects of factors related to inequalities of information society typically by the assistance of diffusion models (e.g. Hüsing et. al, 2001; OECD, 2001), primarily starting from that inequalities are basically determined by the adaptation level of ICT. Social and spatial diffusion in time is characterised by a logistic curve, which shows a time-lagged shape depending on the development level of the analysed target group (Figure 1.). As a result of later adaptation certain social groups (for example peripheral regions) are becoming relatively lagged behind, which can be realised in social and spatial inequalities. In phases of the adaptation process different types of inequalities can be discovered. In early adaptation phase, when only few applies ICT, differences can be seen in accessibility, in the phase of diffusion the differences are present between users and non-users, while in the phase of saturation differences in quality can be emphasised. As a result, ICT-based inequalities can more or less be measured by the society's adaptation level.

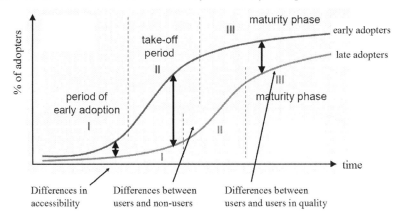

Figure 1. Diffusion model of ICT innovations in the group of early and late adopters (own construction based upon Hüsing et al. 2001)

The theory that ICT-related inequalities have different contents depending on the phase of ICT adaptation were further detailed for example in social (Galácz and Molnár, 2003) or in methodological contexts (Dolni č ar, 2008). Additionally, the above mentioned model is suitable to explain the changing nature of the ICT-related geographical inequalities as well, if regional adaptation level of ICT innovations were taken into account. In this modified model the first order geographical inequalities appear in connection with infrastructural differences in the traditional and "physically existing" space, which can also be named as the external space of the information society and economy (Table 1.). This phase of regional inequalities is basically characterised by differences of accessibility to physical infrastructure. Typically the unequally built-up environment of information and communication infrastructure is among the most important measurable factors. The second phase of ICT-related geographical inequalities has fewer relations with external space; rather it is dependant on internal space inequalities those are disparities of social origin. In this phase accessibility is still an important factor of inequalities, however, with decreasing significance, while usage inequalities are gaining notable importance. In the third phase the main accessibility disparities are disappearing or dissolving and the usage differences remain as inequality motives, however, this time only the quality differences are counting. Technically this phase is the most difficult to measure, since there are only few regional datasets about the quality of ICT usage, however, some experiments already successfully managed to collect data on this topic.

Main character of regional inequalities	Main regional inequality dimensions	Main type of regional inequality factors	Typical phase of appearance in diffusion theories
External inequalities	Regional inequalities of accessibility to physical infrastructure	Traditional	Early
Internal inequalities	Regional inequalities of social origin in accessibility and usage	Transforming	Take-off
Quality-based inequalities	Regional inequalities in the quality of usage	New	Mature

Table 1. Different phases of ICT-based regional inequalities

If we look on recent days' ICT-related regional inequality processes, we might discover that some of the above mentioned dimensions of disparities are rather "old" ones, while others are substantially new. This coincides with our three basic types of inequality factors, since factors, which appear in the early phase of ICT inequalities slowly become traditional ones (by forming the stable background of infrastructure inequalities). The early appearing disparities at last could have a decaying importance. At the same time in the take-off period of ICT diffusion the typical inequality factors obtain a transforming character. Nowadays several factors could be found, which already appeared a couple of years or decades ago, but have somewhat altering significance currently. In the last, mature phase of ICT diffusion the adaptation level differences are decreasing, while completely new inequality features are appearing. Recently the newest inequalities can possibly be connected to some previously unknown dimensions (that are typically related to quality differences). All in all, the regional disparities of the information age can be characterised by both traditional, transforming and new factors of inequalities currently, which could be observed by the evaluation of accessibility and usage parameters.

2. INEQUALITIES OF INFORMATION ACCESSIBILITY

2.1 Has geography got any role in inequalities when internet is "everywhere"?

It seems to be obvious that information accessibility has an increasing importance in modern age inequalities (see e.g. Kim and Kim, 2001), however, this topic is many times explained as aspatial, since internet and other cyber-technologies are available everywhere in the globalised world (Grieg, 2002). On the other hand, there are significant contributions to the concept that the possibilities of information accessibility induce definite differences among locations (Alonso-Villar and Chamorro-Rivas, 2001). Accordingly, the statements should be clarified and the questions have to be answered: why space should be stressfully emphasized in connection with information accessibility? Why is it important to deal with spatial questions in a world, where information – that are the key factors of social and economic development – are available theoretically everywhere?

To be precise, we only think that information is freely accessible for everyone and at all places of the globe. It was many times proved that the role of geographical space could be considered as a borderless and friction free world (Ohmae, 1990; Lewis, 1998) only if we look on the topic as a utopian thing. We could only theoretically state that ensured by new information and communication technologies the everyday troubles originated from spatiality could disappear, namely the ardently wished dream, the overcoming on space may become reality. Empiric results on the other hand still prove that geography matters today as well (Morgan, 2001; de Blij, 2007). This concept realised that previous geographical principles are also standing their ground in recent new environment. It is important that possibilities of information communication network connections and infrastructural grounds of bandwidth, which determine the speed of communication connections, are still unequally distributed in space. This new form of communication is dependent on real world's spatial bounds, on geographical position of access points, materiality of cables, as well as on other infrastructural, social and economic influences outside the world of wires. We should note that no bit can proceed via the Net without passing through kilometres of wires and optical fibres or tons of computer hardware tools, which are all in physical space indeed, and are forming the physical frames of information accessibility.

2.2 Geographical patterns of information accessibility differences

Since there are infrastructural bounds of the chance of getting information, and the pattern of the built-up infrastructure is not equalized spatially, the inequalities should have geographical consequences as well. Telecommunication channels, cable networks of information transference are representing the specific at the same time significantly important material fundamentals of the communication infrastructure that is forming the technical system of conditions of the information society. Actually the most important "public utility" of the information society, the cable system of information transmission plays the main role in the infrastructure-centred version of the external space of the information economy and society.

Concerning regional differences, the level of built up infrastructure as well as distance from access points of networks is assumed to be more unfavourable in geographically peripheral places. Accessibility is though a central category of the geography of information society. It worsens the chance of peripheries since the deployment of technical systems as the "soul" of network society is defined by market regularities of economy, hence infrastructure differentiates society and space also on its own. Centre-periphery relations live further in urban-rural differences, additionally inequalities are defined along city-hierarchy as a result of that nodes of information and communication networks are to be found basically in urban spaces, and the density of connecting services and activities is also the highest at these locations.

To test this assumption, empirical statistical experiments should be carried out by collecting regionally detailed data on information accessibility. As a starting point we analysed the existing methodology to find the best measures of regional inequalities. Although there are many internationally well known attempts to measure ICT-based regional inequalities or at least the level of information society development (see e.g. ITU, 2012), the formulated methods can not be implemented one in one for all kind of regional analysis. The major problem is that international indices take into account variables, which are possible to be collected on country levels, but are rarely available for smaller regional units. The lack of territorially detailed data (basically due to the lack of small scale data collection) resulted that a large number of indicators should be left out from analysis, or alternative solutions should be found. Also due to the novelty of factors, regionalists often struggle with data problems, and can therefore make only general models and measurement experiments of their own; no wide-spread consensual measure or methodology has yet evolved. On regional differences in the development of the information society there are some notable early research experiments in Hungary, which also reflects the new character of this topic (Nagy, 2002).

Since the term of accessibility is a complex one, no simple indicator can be found to characterise it; therefore multi-variable methods have had to be elaborated. To quantify disparities of information accessibility already many experiments were carried out (Corrocher and Ordanini, 2002), mostly dealing with complex sets of indicators, featuring infrastructural and social causes of information accessibility. Typical complex analyses apply indicators that were formerly also important in affecting inequality processes, and on the other hand new indicators that have recently become indispensable. For example, the calculations and the methodology of World Times and International Data Corporation (2001) use 23 different indicators in its complex index. Among the indicators we find those representing the phases of invention, innovation, diffusion and adaptation of the innovation chain. An other widely spread methodology is represented by the International Telecommunication Union's Digital Access Index (DAI), which applies the direct indicators of accessible infrastructure and costs, as well as the indirect indicators of social adaptation. Accordingly, based upon international examples, our calculation – represented in the followings – tried to find the best selection of variables in relation with regional scale information accessibility.

In order to represent the ICT-infrastructure based regional disparities within Hungary, microregional (LAU-1) level data were collected for 174 spatial units. The first dataset was formulated by ICT-infrastructure related indicators, which represent the accessibility of information. Data were provided by the Hungarian Central Statistical Office and by surveys of GKIeNET (an ICT research company in Hungary). The dataset was created for 2010 depending on data availability. The final dataset comprehends the following indicators:

- Number of personal computers in households per 1000 people (Source: GKIeNET)
- Number of mobile phone subscriptions per 1000 people (Source: GKIeNET)
- Number of telephone main lines and ISDN lines per 1000 people (Source: HCSO)
- Number of cable TV subscriptions per 1000 people (Source: HCSO)

The complex index of information accessibility was created by the application of the well known simple Bennett methodology (Bennett 1954). Data were represented as percent of the maximum value and averaged by small regions with the following simple formula (1):

$$I_j = \frac{\sum_{i=1}^{N}\left(\frac{x_{ij}}{x_{i\max}} \cdot 100\right)}{N}$$

(1)

where Ij is the complex index of information accessibility in region j, Xij is the value of indicator i in region j, Ximax is the maximum value of indicator i in the dataset, and N is the number of indicators. Values of the estimation range from 0 (the worst) to 100 (the best).

Figure 2. represents unweighted results of Hungary's regional structure of information accessibility. The map shows the definite difference observed by city-hierarchy, which is reflected by the above average attendance of urban areas. Meanwhile another significant feature is the lagging of the eastern part of the country. Regional differences between eastern and western parts of the country, particularly the lagging of the eastern Great-Plains regions is remarkable. At the same time maximum values of the index are located mostly in the agglomeration of Budapest, in metropolitan regions (Győr, Debrecen, Szeged), as well as in some adjacent zones in Central-Transdanubia and Northwest-Hungary, whilst the minimum values of the index can be connected mostly to small regions of East- and Northeast-Hungary.

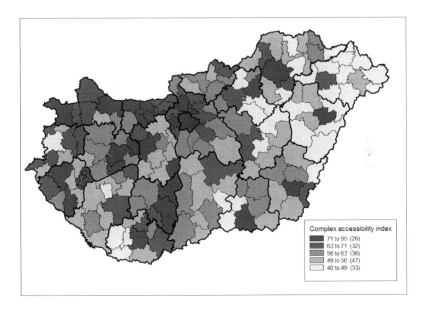

Figure 2. Complex information accessibility index of microregions in Hungary (2010)

The result map mirrors the unequal chance of getting information in different regions of Hungary, although there are some other geographical factors of accessibility, which should be taken into account, too. Besides statistically measurable data also geography-related disparity motives of infrastructure availability could be found in the background. We should be aware that physical geography also determines the possibility of accessing telecommunication channels. There are bounds and barriers of "overall information accessibility" due to constraints of geographical environment, in which for example the relief or the artificial environment could also play an important role (but basically on micro scales, and with minor influences)(Figure 3.). This confirms again that geography still matters in the information world.

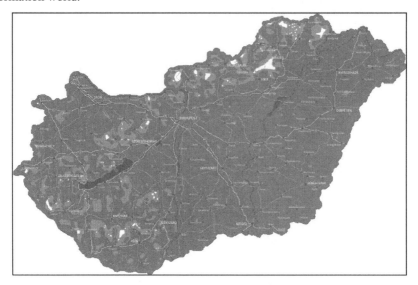

Figure 3. The accessibility of mobile phone networks in Hungary: signal strength map of mobile communication (based on data of www.t-mobile.hu)

3. INEQUALITIES OF INFORMATION USAGE

Geography is important in the information age not only because that material infrastructure of information and communication technologies is unequally distributed in space, but also because there are social intentions to "traditionally" use space even though cyber-technologies make it possible to communicate from any distance. Since the depth of communication interactions is becoming more and more important, it is not only enough to access the information channels, additionally the mode and location of information usage gets an increasing attention.

While technological innovations are continuously diffusing in time, the role of primer ICT background differences in regional inequalities is beginning to decline, since following the typical logistic curve of ICT diffusion. It results that from first order geographical disparities of direct infrastructure accessibility we are stepping towards the increasing importance of second order usage disparities with the revaluation of social, economic and other soft components instead of hard physical factors. Since infrastructure development policies (basically in developed countries) have recognized the necessity of ICT development, increasing number of people have become able to access the new information channels, resulting that accessibility could now be treated sometimes as a background problem and other secondary topics happen to outcrop. In contrast with accessibility differences, recently a new type of disparity emerges: the differences between users in the frequency and way of usage.

This can also be proved by empirical experiments, therefore further statistical data were collected on the level of Hungarian microregions. This time the created dataset was focusing on indicators, which could better reflect usage habits of local people, companies and institutions. Data were provided by surveys of GKIeNET (an ICT research company in Hungary). The final dataset for measuring usage disparities comprehends the following indicators:

- Average level of e-administration (Source: calculations based on GKIeNET)
- Number of internet users per 1000 people (Source: GKIeNET)
- Share of companies with websites (Source: GKIeNET)
- Number of internet subscriptions per 1000 people (Source: GKIeNET)

By applying the same Bennett's methodology this time the unweighted complex regional indices for information usage were determined for each microregion. The outcomes of the calculations (Figure 4.) reflect somewhat similar, but also different spatial structure related to the accessibility map. This time the map shows sharper centre-periphery differences, with best results in the agglomeration zone around Budapest, and observably low values in areas relatively far from the capital. Again, only some urban microregions have better than average indices. The best results could be found in microregions of Central-Transdanubia and in Central-Hungary, while the lowest ones are observable in south-western, eastern and north-eastern areas.

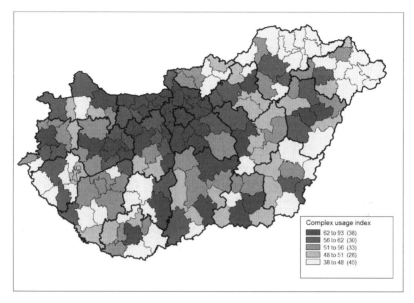

Figure 4. Complex information usage index of microregions in Hungary (2010)

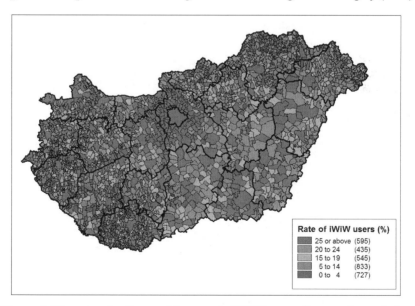

Figure 5. The rate of online social network (iWiW) users from total population in the settlements of Hungary (2008)

The relatively large concentration of the better usage-indices in central parts of the country on Figure 4. assumes that some distance factors are present in the background of information usage results.

This was also confirmed in an other examination, where distance and size effects were tested in regional disparities of using online social network sites (Lengyel and Jakobi, 2012).

Accordingly, if we look on the map, which represents the share of population, who were registered at the largest Hungarian online social network site, it can be observed, that central parts of Hungary, as well as urban zones have usually better, while peripheral and rural areas with smaller settlement size have typically worse user rates (Figure 5). The map was created by using data of iWiW (International Who Is Who), which is a leading online social network provider in Hungary with more than 4 million users. The outcomes of the research certify again that certain components of geography matter in information age inequalities.

4. DISPARITIES OF SPATIAL INFORMATION USAGE – AN EXAMPLE ON QUALITY DIFFERENCES

According to theories about the changing nature of ICT-based inequalities the newest features in connection with regional disparities of information society development should be related to quality issues. By now we have exceeded the initial take-off phase in many respects. The concept emphasises that recently the vast majority of the population has the chance to access information, and also many of them have devices to use new technologies (a lot of people have internet or mobile phone subscriptions), however, only few applies the newest achievements or only few participates the information society actively. In this new type of disparity the quality and not just the quantity counts, which can be possibly observed for example through the usage inequalities of a specific information resource, the spatial information. Spatial (or geographical) information can be considered as qualitatively higher level of information that is also indicated by the ever wider diffusion of applications supporting spatial information usage.

Although spatial information usage is essentially device-independent, yet it is worth to put special emphasis on application possibilities of smartphones. Namely these devices made it possible to easily access spatially sensitive data for those, who had previously not so much ambitions about it. The number of smartphone users recently shows especially dynamic growth worldwide, just like in Hungary. In 2011 statistics reported 15-24% of penetration within the whole population of Hungary (Pintér, 2011), while in 2012 one third (or others assume even higher proportion) of the users have smartphones. Smartphones provide active spatial data usage possibilities through the built in GPS devices, bringing the advantages of GPS technology closer to everyday users. It should be also mentioned in connection with telecommunication trends that the market of separated GPS receivers is shrinking, while the mobile software market (the market of GPS-related smartphone applications) is still expanding. According to Hungarian surveys 22% of the mobile devices have already GPS inside, and more than half of the users reckon the usage of smartphone for positioning as possible in the near future. The survey also reflected on the market potential of location based applications, since almost half of the respondents would pay more or less in order to avail themselves of such services. Territorial level of spatial information usage beyond the direct observation of the diffusion of GPS-enabled devices and other GIS applications can be experimentally determined or at least approximated by other direct or indirect instruments. The frequency of occurrence and the usage of spatial information can be indicated for example by the diffusion level of geocoded information. There are several websites, where published information are provided with clearly defined location parameters (geocodes), resulting that up-to-now spatially independent information also gained spatial attributes. In an HTML environment geocoded information can be placed in the source of website by geotags. By collecting and mapping geotagged data it becomes possible to determine areas, where users publish spatial information more frequently.

Geotagged spatial information can particularly mirror the inequalities of spatial information usage in case of multi-user-edited open websites (such as Twitter or Wikipedia)(see e.g. Graham and Zook 2011). Figure 6. shows the spatial density of geocoded Wikipedia information that can be located within Hungary (due to lack of data the example shows only English and French language results in Hungary). The geographical pattern naturally does not depict the real traffic density, only the relative density of information with spatial content is reflected. The map for the Hungarian results shows again a relative concentration near and inside Budapest but also the wide dispersion of geotags around the country can be noticed.

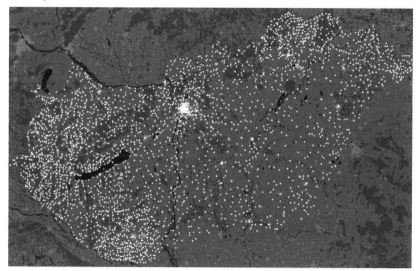

Figure 6. Geographical density of geotagged Wikipedia information in Hungary, 2013
(Source: http://wikiproject.oii.ox.ac.uk)

5. CONCLUDING REMARKS

By understanding that information age inequalities have both older and newer motives, a complex group of factors can be made up in order to get a whole picture on recent days' ICT-related regional inequality processes. Nowadays, both accessibility, usage volume and usage quality disparities are existing, which all should be taken into account in regional information society analysis.

Additionally, geographers have a further task to provide necessary information on understanding current processes of the world. It seems to be needful to shift emphasis on spatial information quality analysis both statistically and empirically, since spatial information is one of the fastest emerging and one of the newest special kind of information, which could induce or perhaps reduce differences between groups of people. There are already several examinations, which reflect that geographical information gained new and dynamic possibilities to reach people by map-crowdsourcing, public participatory GIS or volunteered geographic information solutions (see Gryl, 2012), which all give chance to increase the reputation of spatial information and hereby geography.

6. ACKNOWLEDGEMENTS

Research work of the author was supported by the Bolyai János Scholarship of the Hungarian Academy of Sciences.

REFERENCES

Alonso-Villar, O., Chamorro-Rivas, J-M. 2001. How do producer services affect the location of manufacturing firms? The role of information accessibility. Environment and Planning A: 33 (9): 1621-1642.

Bennett, M. K. 1954. The World's Food. New York, USA: Harper and Row.

Castells, M. 1996. The Rise of the Network Society. The Information Age: economy, society and culture. Oxford, UK: Blackwell Publishers.

Castells, M. 1997. The Power of Identity. The Information Age: economy, society and culture.

Oxford, UK: Blackwell Publishers.

Castells, M. 1998. The Informational City – Information Technology, Economic Restructuring and the Urban-regional Process. Oxford: Basil Blackwell.

Corrocher, N., Ordanini, A. 2002. Measuring the digital divide: a framework for the analysis of cross-country differences. Journal of Information Technology: 17 (1): 9-19.

de Blij, H. 2007. Why Geography Matters. Oxford: Oxford University Press.

Dolničar, V. 2008. Application of an integral methodological approach to measuring the dynamics of the basic digital divide. Observatorio Journal: 4: 065-093

Fichman, R.G. 1992. Information Technology Diffusion: A Review of Empirical Research. Proceedings of the Thirteenth International Conference on Information Systems. Dallas, 195-206.

Galácz, A., Molnár Sz. 2003. A magyarországi információs egyenlőtlenségek. In: Internet.hu – A magyar társadalom digitális gyorsfényképe. ed. Dessewffy, T., Z. Karvalics, L., 138-159. Budapest: Aula Kiadó.

Goddard, J., Gillespie, A., Robinson, J., Thwaites, A. 1985. The impact of new information technology on urban and regional structure in Europe. In: The Regional Economic Impact of Technological Change. ed. Thwaites, A., Oakey, R., 215-242. London: Frances Pinter.

Graham, S., Aurigi, A. 1997. Virtual cities, social polarization, and the crisis in urban public space. Journal of Urban Technology:4 (1): 19-52.

Greig, J.M. 2002. The End of Geography? Globalization, Communications, and Culture in the International System. Journal of Conflict Resolution: 46 (2): 225-243.

Gryl, I. 2012. A web of challenges and opportunities. New research and praxis in geography education in view of current web technologies. European Journal of Geography: 3 (3): 33-43.

Graham, M., Zook, M. 2011. Visualizing Global Cyberscapes: Mapping UserGenerated Placemarks. Journal of Urban Technology: 18 (1): 115–132.

Hüsing, T., Selhofer, H., Korte, W.B. 2001. Measuring the digital divide: A proposal for a new index. IST Conference, 3. Dec. 2001, Düsseldorf.

ITU 2012. Measuring the Information Society. Geneva: International Telecommunication Union.

Jakobi, Á. 2004. Traditional and New Causes of Regional Inequalities in Hungary. In: Emerging Market Economies and European Economic Integration, ed. R.S. Hacker, B. Johansson, C. Karlsson, 160 – 185. Cheltenham: Edward Elgar Publishing.

Kim, M.C., Kim J.K. (2001) Digital Divide: Conceptual Discussions and Prospect. In: The Human Society and the Internet Internet-Related Socio-Economic Issues, ed. Kim W., Ling T.W., Lee Y.J., Park S.S. 78-91. Seoul: First International Conference, Human.Society@Internet 2001 Seoul, Korea, July 4–6, 2001 Proceedings.

Lengyel, B., Jakobi, Á. 2012. The offline landscape of an online social network: distance and size shaping community spread and activity. 52nd European Regional Science Association Congress, Bratislava (Paper 01034.)

Lewis, T.G. 1998. Friction Free Economy: Strategies for Success in a Wired World. New York: Harper Business.

Masuda Y. 1988. The Information Society as a Postindustrialised Society (Az Információs Társadalom mint Posztindusztriális Társadalom). Budapest: OMIKK.

Morgan, K. 2001. The exaggerated death of geography: localised learning, innovation and uneven development. The Future of Innovation Studies Conference, The Eindhoven Centre for Innovation Studies, Eindhoven University of Technology.

Mossberger, K., Tolbert, C.J., Stansbury, M. 2003. Virtual inequality: beyond the digital divide. Washington D.C.: Georgetown University Press.

Nagy, G. 2002. Területi különbségek az információs korszak küszöbén. Területi Statisztika: 42 (1): 3-25.

Norris, P. 2001. Digital Divide: Civic Engagement, Information Poverty, and the Internet Worldwide. Cambridge, UK: Cambridge University Press.

Odendaal, N. 2003. Information and communication technology and local governance: understanding the difference between cities in developed and emerging economies. Computers, Environment

and Urban Systems: 27 (6): 585-607.
OECD 2001. Understanding the Digital Divide. Paris: OECD Publications.
Ohmae, K. 1990. The Borderless World. New York: Harper Business.
Pintér, R. 2011. Az okostelefonok terjedése Magyarországon. Információs Társadalom: 11 (4): 48-63.
Porat, M. U. 1977. The information economy. Definition and measurement. Washington D.C.: US Department of Commerce, Office of Telecommunications.
Servon, L. J. 2002. Bridging the Digital Divide: Technology, Community, and Public Policy. Oxford, UK: Wiley-Blackwell.
Trujillo, M.F. 2001. Diffusion of ICT Innovations for Sustainable Human Development. www.payson.tulane.edu/research.
Umesao, T. 1963. Joho sangyo ron [On Information Industries]. Hoso Asahi: (jan): 4-17.
van Dijk, J. 2005. The Deepening Divide: Inequality in the Information Society. Thousand. Oaks, London, New Delhi: Sage Publications.

DYNAMIC OPPORTUNITY-BASED MULTIPURPOSE ACCESSIBILITY INDICATORS IN CALIFORNIA

Pamela DALAL
University of California, Santa Barbara, Department of Geography,3625 Ellison Hall, UC Santa Barbara, Santa Barbara, CA 93106-4060 ,
dalal@geog.ucsb.edu, http://geog.ucsb.edu/geotrans

Yali CHEN
University of California, Santa Barbara, Department of Geography, 3625 Ellison Hall, UC Santa Barbara, Santa Barbara, CA 93106-4060,
http://geog.ucsb.edu/geotrans, yali@geog.ucsb.edu

Konstadinos G. GOULIAS
University of California, Santa Barbara, Department of Geography, 5706 Ellison Hall, UC Santa Barbara, Santa Barbara, CA 93106-4060
http://geog.ucsb.edu/geotrans, goulias@geog.ucsb.edu

Abstract

Accessibility, defined as the ease (or difficulty) with which activity opportunities can be reached from a given location, can be measured using the cumulative amount of opportunities from an origin within a given amount of travel time. These indicators can be used in regional planning and modeling efforts that aim to integrate land use with travel demand and an attempt should be made to compute at the smallest geographical area. The primary objective of this paper is to illustrate the creation of realistic space-sensitive and time-sensitive fine spatial level accessibility indicators that attempt to track availability of opportunities. These indicators support the development of the Southern California Association of Governments activity-based travel demand forecasting model that aims at a second-by-second and parcel-by-parcel modeling and simulation. They also provide the base information for mapping opportunities of access to fifteen different types of industries at different periods during a day. The indicators and their maps are defined for the entire region using largely available data to show the polycentric structure of the region and to illustrate this as a method that can be applied in other polycentric regions.

Keywords: *hierarchical spatial choice, spatial cluster analysis, multi-scale representation*

1. INTRODUCTION

Recent legislation in California aiming at stricter mobile source emissions control and planning for dramatic decreases in Greenhouse Gas (GHG) emissions emphasizes the need for integrated land use policies with transportation policies. This is expected to happen with planning tools such as a Sustainable Communities Strategy (SCS), which among its many objectives also needs to understand residential location and relocation decisions and explain possible futures under different scenarios of policy to a variety of audiences including decision makers and professional planners and engineers (see http://www.ca-ilg.org/SB375Basics). Similar to many European jurisdictions land use planning is in the local level (e.g., the City) and transportation planning is at higher levels. In the US, transportation planning is a foundational activity of Metropolitan Planning Organizations (MPO) that were created in the 1960s to ensure coordinated planning among local jurisdictions. Two of the most important elements of this planning activity are the Long Range Regional Transportation Plan (LRTP), which every 4-5 years creates a vision and a path to reach goals that protect the environment, foster economic growth, and ensure equity and the second is the development of the regional Transportation Improvement Program, which is an multi-billion USD investment plan to satisfy LRTP goals. The recent legislation adds the land use and transportation goal with a SCS to the MPOs. In California the four largest MPOs (their region surrounds Los Angeles, Sacramento, San Diego, and San Francisco) are also required to build simulation models to assess scenarios for meeting specific targets of GHG emissions by 2020 and 2035. At the heart of these urban simulation models are behavioral equations of residence, workplace, and school location choices by households and their members together with activity and travel behavior equations to represent the daily activities and movements of people in the region. All these models and simulation tools are currently developed and often face two major stumbling blocks: a) lack of understanding of the behavioral processes we try to change with the new policies; b) lack of suitable tools to explain spatio-temporal phenomena that emerge from complex interactions among people. The short statement below is indicative of the relationships we should disentangle, understand, and recreate in predictive urban simulation models.

"A household's decisions of residential location, workplace, activities and travel pattern are an inextricably entangled weave of mutual interdependencies and constraints. Each of these choices is connected to all the others,..." Eliasson, 2010 pg 138. At the core of this we find spatial structure analysis and particularly the spatial structure analysis of urban environments, which is a key informant about location choices of people. This is becoming extremely important in assessing policy actions that change land use to influence travel behavior and attempts to steer it away from using automobiles. The assessment of these policies, in large metropolitan organizations, is done with urban simulation software that employs discrete choice models (see the review by Waddell, 2002). These models predict location choices (e.g., residence, workplace, school, or possibly other major pegs at which activities take place) using as explanatory variables a variety of location attributes in more or less complex forms of accessibility indicators. In this area we see an increasing sophistication of techniques such as multi-scale approaches that account for spatial and behavioral heterogeneity while attempting to solve some problems with spatially correlated explanatory variables and counteract potential fallacies. A sample of recent advances are papers by Bhat and Guo, 2004, Guo and Bhat, 2004 and 2007, Mohammadian et al., 2005, Sivakumar and Bhat, 2007, Sener, Bhat, and Pendyala, 2011. In parallel, in policy oriented circles and among advocates of specific land use actions to change travel demand, we also see growing literature offering an emerging typology of spatial structure indicators (e.g., land use density, diversity of land uses, characteristics of the highway infrastructure, proximity to public transportation). The review and meta-analysis of Ewing and Cervero (2010) provide a comprehensive report of the use of these indicators, a link between the policy literature and an estimate of the impact of these indicators on a limited set of travel indicators. As one would expect spatial structure is very important in studies about the historical evolution of settlements (typical example is the urban growth application of Stanilov and Batty, 2010).

In all these studies we find accessibility to opportunities for employment and/or activity participation explaining the location choices of households. Accessibility indicators can take a variety of forms but they almost always include some measure of location attractiveness (e.g., amount of activity, number of stores, variety) weighted or buffered by measures of impedance (e.g., travel time to reach activities). Location choice is also a function of a variety of other factors including cost (e.g., price of homes), spatial ethnic segregation (e.g., immigrant ethnic enclaves or other social processes that motivate people of similar cultural traits to co-locate), social exclusion (e.g., specific groups may be excluded from different areas of an urban environment by policy or tradition), or temporary co-location for education or other reasons (e.g., attending a specific college, serving in the military).

Underlying all this, a hierarchy exists in the spatial organization of opportunities that characterizes each living environment. In fact, large urban environments are no longer monocentric but show clear emergence of polycentrism (Giuliano and Small, 1991, and the review by Anas et al., 1998). It is important then to identify and describe underlying spatial structures (Hughes, 1993) but this is not a trivial task and should account for transportation infrastructure. Hierarchies are based on geographic space (e.g., region, city, neighborhood, city-block, land parcel), time (e.g., historical time, day of the week, time of day), in-situ social networks (e.g., ethnicity, religious meeting places), and type of activity opportunities (e.g., retail, arts and entertainment, leisure).

To the best of our knowledge a method that recognizes this hierarchy explicitly and provides classification of locations using accessibility and segregation, as well as, employs informative opportunity indicators does not exist or it is done in a somewhat ad-hoc opportunistic way. A more systematic approach is useful in developing choice sets, creating new type of explanatory variables for discrete choice models, estimating models tailored to localities of special character, and helps us characterize spatial structure and its evolution for urban simulation models. It may also help us understand, explain, and support the input to and output from urban simulation models.

To partially fill this gap we report in this paper findings from a pilot research project with focus on the Southern California five county region (which is also the largest MPO in California) that takes advantage of accessibility indicators computed at a fine level of spatial resolution (the US Census block), for fifteen types of employment, and different times during a day (accounting for opening and closing of businesses and the presence of congestion in different parts of the region at different times of a day). We use these indicators to develop spatial clusters that are able to classify each block based on its own intensity of activity availability and the opportunities available at its adjacent (contiguous) blocks. In this way we derive spatial clusters for a selection of activity opportunities for each block by first computing this value from each land parcel within the block. These spatial clustering indicators are then used in another clustering process, using Latent Class Cluster Analysis, to classify different parts of Southern California into five categories that range from high accessibility, to medium accessibility, and finally to very low accessibility. We repeat this using the same method for four time periods of a day to describe the evolution of the region during a day and how different localities change in their ability to provide services to their residents. We also study these different groups of blocks in terms of their resident characteristics.

In the next section we describe the accessibility indicators used here followed by a section on the spatial cluster analysis and the Latent Class Cluster Analysis together with a description of the time-of-day dynamics. Then we review the findings in the correlation between spatial structure and resident characteristics. The paper concludes with a summary of findings and next steps.

2. ACCESSIBILITY INDICATORS AND THE REGION

The block-level accessibility measures described in Chen et al., 2011 are the main source of information to describe the spatial structure of Southern California. These measures are computed at a fine level of spatial disaggregation, which is the US Census block. As described in Chen et al., 2011, we used multiple databases to describe the opportunities available at different levels of spatial aggregation and converted all data into information for each block while rectifying any missing or miscoded information through comparisons of different sources of data.

The end result is an account of the number of persons at each block working in any of the fifteen different and mutually exclusive industry types that are: a) Agriculture, forestry, fishing and hunting and mining; b) Construction; c) Manufacturing; d) Wholesale trade; e) Retail trade; f) Transportation and warehousing and utilities; g) Information; h) Finance, insurance, real estate and rental and leasing; i) Professional, scientific, management, administrative, and waste management services; j) Educational; k) Health; l) Arts, entertainment, recreation, accommodation and food services; m) Other services (except public administration); o) Public administration; p) Armed forces. All blocks are connected to a roadway network that includes for each of its links estimated speed for different periods of a day.

This network is used to compute shortest paths among all the blocks. There are approximately 203,000 blocks that cover the entire (mega)region surrounding Los Angeles (called the Southern California Association of Governments region). Using these shortest paths we identify for each block all the other blocks that are within 10, 20, and 50 minutes to create travel time buffers. Then, for each period in a day and for each of the fifteen industries we count the number of persons employed by each industry type within each buffer.

To account for the different opening and closing times of activity opportunities the number of employees that are reachable in the accessibility indicators changes for each hour in a day. To derive this time of day profile we use information of arrivals and departures from work sites available in a travel survey. In this way the resulting accessibility indicators change with space and time to reflect the rhythms of activity in the region for which they are defined. More details about the method are reported in Chen et al., 2011 and an application to neighborhood analysis in Dalal and Goulias, 2011. Due to the region's size and vastly varied land use, population for any given block can range from zero to over 7,000 residents. However, the attributes of the built environment described are not equally distributed over the county. For example, the density of transportation infrastructure increases with population density. This creates different levels of network connectivity and accessibility between urban and rural neighborhoods (Chen, et al, 2011). Other attributes include the distribution of opportunities, which are often more dense in more urban areas of Southern California (Dalal and Goulias, 2011).

In this analysis we use accessibility indicators as the core material in developing groups of similar spatial structure. We selected to work with four periods in a day that follow the current SCAG four-step model that provides travel speed and time for each roadway segment for four time periods, AM peak (6 AM to 9 AM), PM peak (3 PM to 7 PM), Midday off-peak hours (9 AM to 3 PM), and Nighttime off-peak hours (7 PM to 6 AM). This allows the calculation of shortest path travel time between blocks for each of these four different periods in a day and analyze blocks in terms of the 10 minute accessibility indicators of all fifteen types of industries. In this way we have fifteen continuous variables for each of the 203,000 blocks covering the entire Southern California.

Similarly, the population of Southern California is as varied as the landscape. Within the study region, there exists a certain amount of social and demographic stratification. The average density, age, household income, and household size differ between sub-regions and between neighborhoods as shown in Table 1. One example is in race/ethnicity, which can be very diverse, as in Pasadena, or very homogenous, as in Boyle Heights. Thus, while the built environment varies over space, persons with different socio-demographics populate different spaces.

Region	South Bay		San Gabriel Valley		Westside	Southeast	Eastside
Neighborhood	Rancho Palos Verdes	Ingle-wood	Pasadena	Glendale	Santa Monica	Compton	Boyle Heights
Persons/mi	3,084	12,330	5,366	6,368	9,817	9,199	14,229
Median hh size	2.7	3	2.5	2.7	2	3.9	3.8
Median hh income	128,321	46,574	62,825	57,112	69,013	43,157	32,253
Rented housing	18.1	63.6	54	61.6	70.2	43	75.9
Single parent hh	4.8	26.5	13.4	9.4	12.8	22	21.2
Median age	44	29	34	37	38	24	25
% White	62.8	4	39.1	54.1	71.3	1	2
% Latino	5.6	46	33.3	19.6	13.5	57	94
% Black	2.1	46.4	13.9	1	3.5	40	0.9
% Asian	25.2	1.1	10	16.3	7.1	1	2.4

Table 1. Socio-demographics of selected neighborhoods in Southern California; Source: US Census, 2000

In our study, we consider the anisotropic spread of environmental attributes and population characteristics as an underlying preference for choice for residences, workplaces, and other activity destinations. This is accomplished by first accounting for the variation in opportunities and persons over space through spatial clustering methods. Second, the clusters of opportunities are associated with persons to describe a probable set of choices based on household survey data.

3. G* ANALYSIS

This study examines spatial clusters of opportunity access aiming at representing subregions within Southern California that display spatial homogeneity, such as a neighborhood of blocks with high accessibility to locations that offer arts opportunities. We account for spatial heterogeneity from large-scale regional effects by using a local indicator of clustering. In addition, we consider spatial dependency in our data. For any spatial outcome, such as land prices, the values in one location are more influenced by the values of nearby locations than values of far locations. However, it is unlikely that any urban phenomenon is spatially independent, so our approach is to describe spatial dependency through spatial clusters.

Spatial clusters are built on the concept of spatial dependence, in which things closer to each other are more related than things farther apart (Tobler, 1970). A cluster measures the concentration (or dispersion) of values over space, a simple measure of positive spatial dependence or autocorrelation. In our analysis, a positive cluster specifies when a block which is surrounded by more similar blocks than expected at random. Conversely, a negative cluster specifies when a block is surrounded by dissimilar blocks, more than expected at random, indicating negative spatial autocorrelation. Thus, our first step is to measure the concentration or dispersion of urban attributes by locating and defining the spatial extent of spatial clusters (Jacquez, 2008).

However, it is important to note that urban processes, including the distribution of activities, are widely considered to be spatially heterogeneous, in that the outcomes of the processes vary over space (Anselin, 1995; Paez and Scott, 2004; Jacquez, 2008). Spatial heterogeneity is especially true for extensive study areas where large-scale regional effects can influence the mean and variance of spatial processes, thereby biasing spatial clustering analysis of values (Miller, 1999; Buliung and Kanaroglou, 2007).

In our study of Southern California, large variation can be found in the number and density of opportunities available to blocks throughout the day (Figure 1 shows box-plots of the sum of all fifteen accessibility indicators). High-density areas such as downtown Los Angeles would affect estimation in a global cluster analysis, such as Moran's I, thus the use of a local indicator of clustering is highly relevant in a large study area like Southern California. By using a local measure of spatial association, we are able to estimate clusters without large-scale regional bias, and show significant local clusters in rural or outskirt areas (Anselin, 1995).

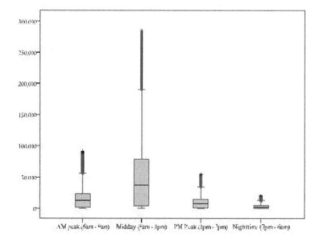

Figure 1. Total opportunities during different time periods for Southern California study regions

In our spatial cluster analysis, we use the G statistic, developed by Getis and Ord (1992), to quantify local spatial autocorrelation and dependency and reveal block-level. This measure was selected for its attractive properties and based on a preliminary pilot study using Los Angeles county data alone. The output of the repeated application of G* to each of the fifteen accessibility indicators are fifteen continuous variables of z-scored spatial cluster accessibility indicators. These are in turn used as criteria variables to identify latent classes (LC) using Latent GOLD® 4.5 software (Vermunt and Magdison, 2005), which is a model-based latent class cluster model building system. The analysis starts with a 1-dimensional LC baseline model (one cluster), followed by fitting successively LC models by adding one dimension (additional cluster) each time. The goodness of model fit is assessed using the Bayesian Information Criterion (BIC) value which is a penalizing statistic for excessive estimated parameters and it is a function of the log-likelihood statistic (LL). The final cluster spatial structure is defined as the model with a low BIC value and less parameters. Taking into account parsimony and model fit statistics a 5-class latent cluster model is estimated with the accessibility indicators for each time of day.

Figure 2 shows the mean (in z-scores) spatial clustering values of industries for the five clusters and Table 2 shows the cluster sizes during the AM, MD, PM, and NT. The largest cluster is Cluster 1 (Crimson), which accounts for a third of all blocks in the study region. Cluster 1 shows high accessibility for all industry types except agriculture and armed forces. It is lowest during the night time which may be explained by store operating hours. Cluster 2 (Orange) is the second largest class and shows an average accessibility to all industry types. The third largest class is Cluster 3 (Gold) and has lowered access to all industries over all times of day. Cluster 4 (Cyan) experiences very low access to all industries and is fourth in size. This cluster is largest during midday, but then shrinks during the PM peak. The smallest cluster is Cluster 5 (Blue) which experiences very low accessibility to all industries except agriculture and armed forces. This cluster expands during the night time and may include blocks that have higher accessibility in other time periods. This outcome shows the temporal variability of accessibility that is captured by the LCCA outcome.

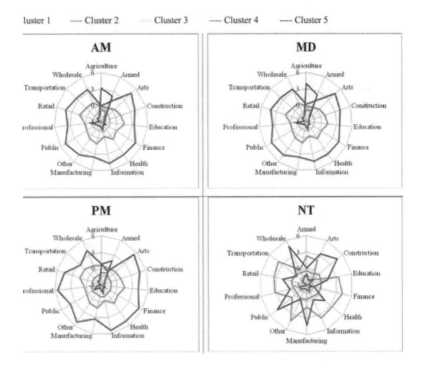

Figure 2. Latent Class Cluster Characteristics

Cluster Size	Cluster 1	Cluster 2	Cluster 3	Cluster 4	Cluster 5
AM	35.84	20.12	18.13	16.07	9.83
MD	33.1	22.7	18.3	18.3	7.6
PM	36.07	21.86	18.5	13.06	10.5
NT	29.3	20.81	18.4	17.76	13.73

Table 2. Size of Latent Class Clusters for time of day

In Figure 3 a map of the blocks classified by cluster membership shows a much clearer spatial structure. There is a clearly definable region in Southern California of high accessibility that stretches along the Pacific Ocean and east to west to the center of the city of Los Angeles. Accessibility is in fact enhanced by the presence of freeways as backbone to this structure. In addition, these maps show the polycentric/multimodal character of the spatial organization of the SCAG region (recall this region includes 190 cities). The four maps together also show that the role of these centers (nodes) is for some centers dynamically homogeneous and for some other centers heterogeneous. For example, the coast has many centers that provide high accessibility throughout the day and even at night time (lower right hand map) it maintains good accessibility but for a smaller number of industries than in the day time. Similarly, some areas in the far northeast of the region have poor accessibility all the time. In contrast, there are groups of blocks (e.g., North Orange County) that have dramatic changes of accessibility at night.

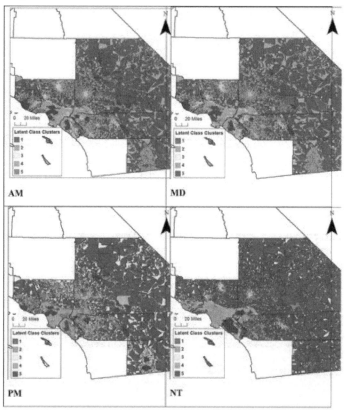

Figure 3. Map of Latent Class Cluster Output for Time of Day

Another reason that clusters show significantly different patterns of accessibility during the night time is also the inclusion of a large portion of zero opportunities for industries at this time, which can bias the clustering outcomes. This is likely why the clusters show a major shift in pattern during the NT, as shown in Figure 3. Cluster 1 is found where clusters of high opportunities are found, with the exception of armed forces and agricultural industries. Cluster 1 is found in the downtown Los Angeles region and a few locations in the east. When comparing the time periods, Cluster 1 is more pronounced in the east in the PM peak, though public and manufacturing opportunities decline. Cluster 2 experiences more heterogeneity in opportunity access with neither high clustering nor high dispersion. In space, Cluster 2 is largely found in the suburban Central Valley extending until the eastern deserts. Importantly, Cluster 2 represents the average heterogeneity in the distribution of opportunities based on the local and global mean in the study region. Cluster 3 shows low levels of dispersed opportunities and is found in small pockets consistent with locations of outer rural towns. Cluster 4 is found in clusters of high dispersion, seen in the Central Valley exurbs and along outer transportation networks. Similarly, Cluster 5 experiences high dispersion, with the exception of high clustering of armed and agricultural opportunities.

Near the coast, Cluster 1 clearly represents mixed land use in high density urban areas. However, Cluster 1 also represents high clustering, or rural centers, when found near areas of dispersed armed and agricultural lands in the east. The dynamics in accessibility are more pronounced in these rural areas, with more high clusters in the AM and PM peak than during work hours. Next, Cluster 3 experiences much more dispersion in the PM peak than in earlier time periods. Translated on a map, the exurbs of Los Angeles Valley lose access to many opportunities during the PM peak possibly from congestion or early store closing hours. This is in contrast to Cluster 4 in rural areas which show no change in low clustering values.

These clusters can be related to socio-demographic characteristics of the resident populations. As shown in Figure 4, high accessibility clusters have the highest level of percent renters, black, asian, and foreign born residents (which is an indication of the different immigration waves in this area). In contrast, low accessibility clusters are higher in home owners living in areas with high levels of vacant housing. The contrast between these two clusters suggests high accessibility may not be an immediate proxy for desirability, and low accessibility may not describe undesirability. In fact, we find an 11.31% to 13.80% of poverty level residents in all clusters with the highest percentage in the two best access clusters. Interestingly, Cluster 3 with low levels of dispersion has the highest percentage of home owners and lowest minority and poverty rates. These are the areas in the outer rim of Los Angeles which experience a decrease in accessibility into the PM peak. Cluster 3 may describe an older, white population that has moved to the exurbs far away from the urban core and then services followed them but did not reach the same high levels as along the coast.

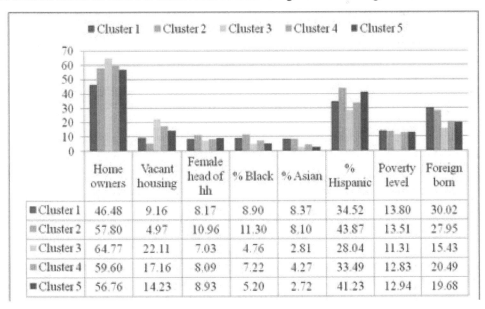

	Home owners	Vacant housing	Female head of hh	% Black	% Asian	% Hispanic	Poverty level	Foreign born
Cluster 1	46.48	9.16	8.17	8.90	8.37	34.52	13.80	30.02
Cluster 2	57.80	4.97	10.96	11.30	8.10	43.87	13.51	27.95
Cluster 3	64.77	22.11	7.03	4.76	2.81	28.04	11.31	15.43
Cluster 4	59.60	17.16	8.09	7.22	4.27	33.49	12.83	20.49
Cluster 5	56.76	14.23	8.93	5.20	2.72	41.23	12.94	19.68

Figure 4. Cluster Socio-demographic characteristics

4. SUMMARY AND CONCLUSIONS

In this paper we present one way to analyze complex spatial systems enhancing our understanding of the dynamics of polycentric cities and of residential, job, and school location choice processes. We first classify a large region into a multilevel nested spatial structure using information from regional models, US Census blocks, and networks. We also employ multiple sets of indicators including resident population and its characteristics, infrastructure provision, activity opportunities by type of opportunities, and housing supply, and synoptic measures of activity and travel behavior. We also introduce time of day as a fundamental element in classifying spatial units at different observation levels. The primary objective of this paper, to illustrate the creation of realistic space-sensitive and time-sensitive fine spatial level accessibility indicators that attempt to track availability of opportunities, is met using largely available data. These indicators support the development of the Southern California Association of Governments activity-based travel demand forecasting model that aims at a second-by-second and parcel-by-parcel modeling and simulation. They also provide the base information for mapping opportunities of access to fifteen different types of industries at different periods during a day reflecting clearly the polycentric nature of the urban landscape of this region. To accomplish this task classification of units within each level is performed in two distinct ways: a) cluster analysis using observed variables to derive groups and their taxonomy; and b) cluster analysis using a set of latent constructs to derive groups and their taxonomy. In our spatial analysis, we include the spatial dependency and homogeneity attributes as latent descriptors of the environment, which are then used to classify Southern California into block clusters. The method used and our findings improve center identification techniques of the type found in McMillen (2001) and Redfearn and Giuliano (2008).

The methods here reduced a complex spatial environment into groups that share similar attributes but also share spatial patterns and provided new insights. What is uncertain is the appropriate scale of the spatial analysis, as the scale of our spatial units and the scale of neighborhood definitions. In our analysis, we use the Census block which varies in size over Southern California. To compensate for resident population size, urban blocks are quite small while parks and rural areas are much larger. Smaller blocks often have more neighbors within a given distance. Likewise, the shape of rural blocks is rarely regular resulting in more contiguous neighbors. The size and shape affect the value of the local mean derived from 'neighboring' blocks of neighbors in the spatial clustering analysis. This is a classic geographic concern known as the Modifiable Unit Problem (MAUP) in which spatial analysis is inherently biased by the data. The MAUP effect is most clearly seen during the nighttime when most blocks report no open businesses. The nighttime clusters are unusual in their size, spatial distribution, and industry characteristics. Interpretation is aided with more descriptive statistics on block-specific attributes like socio-demographics. We need to expand the research reported here and test the use of land parcels as spatial units to reduce the bias from MAUP.

However, when applied as a way to describe the environment and identify subcenters of activity during multiple time points, we find spatial structures related to the hierarchy of transportation networks and industry agglomeration. The next steps in continued research are to enhance the inventory of land use, expand travel modes to transit, account for public lands, and add other indicators that describe the built environment. We also used the spatial indicators in a multi-equation study addressing the relationship between travel behavior and land use patterns using a Structural Equations Modeling framework that is an ongoing study to compare location choices and travel behavior choices among residents of different regions such as the Lisbon metropolitan area in Portugal, The Montreal region in Canada and the Los Angeles region in Southern California (de Abreu, Goulias, and Dalal, 2011). While the context of this study is in Southern California, we believe the method can be extended outside the United States. Similar to Southern California, the geographic region of the European Union contains many major and minor cities, creating a polycentric and complex built environment. Examples include the RhineRuhr, Randstad, Central Belgium, as well as the regions surrounding Paris, Zurich, and London (Taylor, Evans, and Pain, 2006). In fact, this analysis parallels European studies such as POLYNET (http://www.polynet.org.uk) and could support the data analysis in the ESPON 2013 Programme.

Aided by improved transportation networks and relaxed border controls, increased connectivity throughout the region results in a continuous but hierarchical landscape in which opportunity accessibility is likely found at multiple spatial scales. Using the spatial clustering method described in this study, insights into the local, regional, and continental organization of the environment can inform policy decisions regarding the continued integration and collaboration of multiple international entities.

ACKNOWLEDGMENTS

Funding for this research was provided by The Southern California Association of Governments, The University of California Transportation Center (funded by the US Department of Transportation and the California Department of Transportation) , and the University of California Office of the President (Multicampus Research Program Initiative on Sustainable Transportation and the US Lab Fees Program). This paper does not constitute a policy or regulation of any Local, State, or Federal agency.

REFERENCES

Anas, A., Arnott, R. and Small, K. 1998. Urban spatial structure. Journal of Economic Literature: 36: 1426-1464.

Anselin, L. 1995. Local indicators of spatial association-LISA. Geographical Analysis: 27 (2): 93-115.

Bhat, C. R., Guo, J.Y. 2004. A Mixed Spatially Correlated Logit Model: Formulation and Application to Residential Choice Modeling. Transportation Research Part B: 38 (2): 147-168.

Buliung, R. N., Kanaroglou, P. S. 2007. Activity–Travel Behaviour Research: Conceptual Issues, State of the Art, and Emerging Perspectives on Behavioural Analysis and Simulation Modelling. Transport Reviews: 27(2): 151-187.

Chen, Y., Ravulaparthy, S., Deutsch, K., Dalal, P., Yoon, S.Y., Lei, T., Goulias, K.G., Pendyala, R.M., Bhat, C.R. and Hu, H-H. 2011. Development of Opportunity-based Accessibility Indicators. Transportation Research Record: Journal of the Transportation Research Board (in press).

Dalal, P., Goulias, K.G. 2011. Geovisualization of Opportunity Accessibility in Southern California: an exploration of spatial distribution patterns using geographic information systems for equity analysis, GEOTRANS Technical paper, Department of Geography, University of California-Santa Barbara. Accepted for presentation at the 90th Annual Transportation Research Board Meeting, Washington D.C., January 23-27.

de Abreu e Silva, J., Goulias, K.G. and Dalal, P. 2011. A structural Equations Model of Land Use Patterns, Location Choice, and Travel Behavior in Southern California. Paper 12-3422 to be presented at the 91st Annual Meeting of the Transportation Research Board, Washington, D.C., January 22-26.

Doling, J. 1975. The Family Life Cycle and Housing Choice. Urban Studies: 13: 55-58.

Eliasson, J. 2010. The Influence of Accessibility on Residential Location. In Residential Location Choice: Models and Applications, eds. F. Pagliara, J. Preston, and J. Simmonds, 137-164. Berlin: Springer.

Ewing, R., Cervero, R. 2010. Travel Behavior and Built Environment. Journal of American Planning and Association: 76 (3): 265-294.

Getis, A., Ord, J. 1992. The analysis of spatial autocorrelation by use of distance statistics. Geographical Analysis: 24 (3): 189-206.

Giuliano, G., Small, K. A. 1991. Subcenters in the Los Angeles Region, Regional Science and Urban Economics: 21(2): 163-182.

Guo, J. Y., Bhat, C.R. 2004. Modifiable Areal Units: Problem or Perception in Modeling of

Residential Location Choice? Transportation Research Record: Journal of the Transportation Research Board: 1898: 138-147.

Guo, J. Y., Bhat,. C. R. 2007. Operationalizing the Concept of Neighborhood: Application to Residential Location Choice Analysis. Journal of Transport Geography: 15 (1): 31-45.

Hughes, H. L. 1993. Metropolitan Structure and the Suburban Heirarchy. American Sociological Review: 58 (3): 417-433.

Jacquez, G. M. 2008. Spatial cluster analysis. The Handbook of Geographic Information Science, eds. S. Fotheringham and J. Wilson, 395-416. Oxford: Blackwell Publishing.

Litman, T. 2002. Evaluating transportation equity. World Transport Policy & Practice: 8 (2): 50-65.

McMillen, D. P. 2001. Nonparametric Employment Subcenter Identication. Journal of Urban Economics: 50 (3): 448-473.

Miller, H. 1999. Potential contributions of spatial analysis to geographic information systems for transportation (GIS-T). Geographical Analysis: 31 (4): 373-399.

Mohammadian A., Haider, M. and Kanaroglou, P. S. 2005. Incorporating Spatial Dependencies in Random Parameter Discrete Choice Models. Paper submitted for presentation at the 84th Annual Transportation Research Board Meeting, January 23-27, Washington D.C.

Páez, A., Scott, D. 2004. Spatial statistics for urban analysis: A review of techniques with examples. GeoJournal: 61 (1): 53-67.

Pagliara, F., Timmermans, H. J. P. 2009. Choice set generation in spatial contexts: a review. Transportation Letters: 1 (3): 181-196.

Redfearn C., Giuliano, G. 2008. Network Accessibility and the Evolution of Urban Employment. METRANS Project 06-16, Draft Report, University of Southern California, Los Angeles, CA.

Sener I., Pendyala, R. and Bhat, C.R. 2011. Accommodating Spatial Correlation Across Choice Alternatives in Discrete Choice Models: An Application to Modeling Residential Location Choice Behavior. Journal of Transport Geography: 19: 294-303.

Sivakumar, A., Bhat, C. R. 2007. A Comprehensive, Unified, Framework for Analyzing Spatial Location Choice. Transportation Research Record: Journal of the Transportation Research Board: 2003: 103-111.

Stanilov K., Batty, M. 2010. Exploring the Historical Determinants of Urban Growth through Cellular Automata. UCL – CASA paper. http://www.casa.ucl.ac.uk/working_papers/paper157.pdf

Taylor, P. J., Evans, D.M. and Pain. K. 2006. The Organisation of Europolis: Corporate Structures and Networks, The Polycentric Metropolis, 53-64. London: Earthscan.

Thill, J. C. 1992. Choice set formation for destination choice modelling, Progress in Human Geography: 16 (3): 361-382.

Tobler, W. 1970. A computer movie simulating urban growth in the Detroit region. Economic Geography: 46: 234-240.

Van Gent, W. P. C. 2010. Housing Context and Social Transformation Strategies in Neighborhood Regeneration in Western European Cities. International Journal of Housing Policy: 10 (1): 63–87.

Van Kempen, R., Ozuekren, A. 1998. Ethnic Segregation in Cities: New Forms and Explanations in a Dynamic World. Urban Studies: 35 (10): 1631–1656.

Vermunt, J. K., Magidson, J. 2002. Latent class cluster analysis. In Applied Latent Class Analysis, eds. Hagenaars, J.A., McCutcheon, A.L. , 89-106. Cambridge: Cambridge University Press.

Vermunt, J. K., Magdison, J. 2005. Technical Guide for Latent GOLD 4.0: Basic and Advanced. Belmont Massachusetts: Statistical Innovations Inc.

Waddell, P. 2002. UrbanSim: Modeling Urban Development for Land Use, Transportation and Environmental Planning. Journal of the American Planning Association: 68 (3): 297-314.

REDEVELOPING THE GREYFIELDS WITH ENVISION: USING PARTICIPATORY SUPPORT SYSTEMS TO REDUCE URBAN SPRAWL IN AUSTRALIA

Stephen GLACKIN

Swinburne University of Technology, Institute for Social Research.
EW Building, Mail H53, PO box 218, Hawthorn, Victoria, 3122, Australia.
http://www.swinburne.edu.au/, sglackin@swin.edu.au

Abstract

Given the recent publications from Australian State governments demanding greater community and stakeholder engagement in urban planning, as well as calls from international agencies for a reduction in the footprint, and increase in the sustainable planning, of cities, there is now the potential for the advances made in geo-tools to have considerable effect. Arising out of 'Greening the Greyfields', a federally funded, inter-state project examining the feasibility of redevelopment in the middle suburbs, ENVISION was produced as a GIS-based, Participatory Support System, for engaging with the diverse array of stakeholders involved in urban redevelopment. This system was designed to bring wide-ranging land, demographic and market data together to highlight the redevelopment options, and identify potential redevelopment precincts, across metropolitan centres, with the aim of initiating debate between those involved on how best to manage urban growth. The result of this project has seen ENVISION being used at a state and municipal level, where workshops based on its use have begun to highlight the barriers to redevelopment as well as the ways forward for more sustainable redevelopment in the urban Greyfields (middle suburbs with high levels of un-planned redevelopment, high incidences of culturally and technologically obsolete dwellings, on land that is highly undercapitalised). Based on the communicative and deliberative models of community engagement, ENVISION has shown that geo-tools can have considerable affect in the mutual education of stakeholders, in extracting the pertinent issues and potential barriers to redevelopment, and in encouraging groups of experts to produce novel solutions to 'wicked' problems that they could not, without the collaboration that the tool demands, resolve on their own. Ultimately this project highlights the ability of GIS to not only provide an interface to real-time data manipulation, but its power to be used as a tool for communicative education between the diverse perspectives within a politically, technologically, financially and culturally sensitive area.

Keywords: *GIS, participatory Support System, urban redevelopment, education, engagement*

1. INTRODUCTION

The planned adoption of significant levels of stakeholder engagement across state planning strategies in Australia (COAG Reform Council 2012; Department of Planning and Community Development 2012; NSW Government 2012) has highlighted the importance of multi-level interaction to the process of urban planning. Implicit in this engagement is the necessity of collaborative, communicative and deliberative processes where negotiation and interaction between diverse stakeholders drives the envisioning of mutually beneficial futures. This turn towards 'bottom-up', or rather the meeting of bottom-up and top-down planning (Russell 2011), has arisen from a number of areas. The change in governance from centralised structures to localised networks (Gallent and Robinson 2012; Geddes

2006), a growth in community participation methodologies (Creighton 2005; Hartz-Karp 2005; Ramasubramanian 2010; Sanoff 2000; Walters 2007; Wates 2000), critical views on the traditional 'top-down' approaches of planning (Brody et al. 2003; Innes and Booher 2011; Lange 2011; Murayama 2008), and significant success in projects that utilise long-term engagement strategies (Kelly 2010) have largely been the drivers behind this change. These factors, combined with the new multi-disciplinary approaches to solving 'wicked problems' (Roberts 2000), have produced a planning environment where the skills of all stakeholders, as well as the knowledge of multiple perspectives (both expert and local), is required to imaginatively and collectively resolve the complex and divisive issues that arise out of attempting to develop effective urban planning schemes.

In support of this shift, the work of (Newton et al. 2011) has illustrated that in order to effectively capitalise on the redevelopment potential of urban areas, there is a need for consultation across the range of stakeholder groups linked to this process. In particular, they highlighted the need for a platform capable of engaging with building developers, government institutions, community members and the range of experts involved in turning the visions of redevelopment schemes into reality. ENVISION was created to achieve this, where, by obtaining data from a wide variety of sources, the redevelopment potential of urban precincts could be queried and, in redevelopment workshops, the diverse parties could potentially reach agreement on the futures of locales. The tool was also designed as a way to extract the tacit knowledge of experts where, as individuals used the tool, they would reveal the limitations and ways forward for specific redevelopment projects. As such it was designed explicitly for engagement, as both a way to encourage interaction and as a mechanism for identifying the various positions and perspectives within the redevelopment arena, with the aim of transferring this knowledge amongst stakeholders and developing institutional mechanisms for more advanced and sustainable urban redevelopment.

In the context of this paper, the importance of this form of engagement comes from calls for more compact cities (OECD 2012), on the basis that current urban expansion is unsustainable and, by using more sophisticated design and technology, there is currently the potential for far more efficient urban design and redevelopment. Newton (2010) identified the potential of the Greyfield (middle suburb areas with dwellings that are culturally and technologically obsolete) to fulfil this role. A report by Newton, Murray et al. (2010) for the Australian Housing and Urban Research Institute (AHURI) revealed that the current nationwide focus on activity centres (areas of high cultural and economic activity) and transport corridors (areas of high transport and economic activity) as the designated strategic areas to drive urban redevelopment were actually having little effect, with the majority of redevelopment occurring sporadically in Greyfields. However, rather than the full capability of Greyfield redevelopment being realised, it was producing low density typologies in an unplanned and non-strategic fashion. The report further argued that with greater integration of stakeholders (business leaders, government, community members and so forth), the factors preventing more advanced design (zoning, land amalgamation, community concerns) could be overcome.

This research led to Greening the Greyfields, a four-year, federally funded project aimed at implementing the four phases of urban redevelopment: proving the economic viability of agglomeration; identifying precincts and extracting tacit knowledge; visualising redevelopment and developing sustainability metrics; and community engagement. The work that led to the creation of ENVISION came from the second module – precinct identification and extraction of tacit knowledge. What this paper will illustrate is the power of geography and GIS, in the form of a decision and participatory support system (ENVISION), to inform and educate a diverse array of stakeholders. Briefly examining the current state of engagement in planning it will use ENVISION as an example of how community engagement techniques can be built into software interfaces (as well as being included in part of their design) to provide a common platform on which the often divergent voices can manipulate data and ultimately produce models that inform other stakeholders, allowing the many voices to come to consensus around central issues.

2. THE SIGNIFICANCE OF ENGAGEMENT WITHIN DIFFERENT DISCOURSES

The above mentioned strategic planning documents, as well as the growth of policies referring to engagement (Fritze et al. 2009; Herriman 2011; Jarvis et al. 2012; King and Cruickshank 2012; Lawson and Kearns 2010; Le Dantec 2012; Reddel and Woolcock 2004), illustrate the move towards a more decentralised form of decision making, or the move from government to governance. What this process refers to is the gradual movement from centralised authority, or departmental creation of policy, to the localised and issue-specific formation of policies designed to more efficiently resolve the issues of local communities at a local level. Argued variously as an increase in democratic process (Aulich 2009; Gallent and Robinson 2012; Sirianni 2008; Smyth et al. 2005; Sorensen and Torfing 2007) and as encroaching neo-liberalism where responsibilities and costs are placed onto the community (Mowbray 2005), the result of this process is an increase in engagement, support for networks of governance involving multiple stakeholders and arguably more control of local affairs by local agencies. This is essentially the institutionalised aspect of engagement where policies, combined with the changing structure of government, have produced a norm of decentralised collaboration as a way to resolve politically sensitive issues and drive effective subsidiarity (Carson 2011).

Planning has followed this trend towards higher levels of engagement, though from a more critical and pragmatic orientation. Beginning in the mid-1960s, and spearheaded by the social justice movements and discussions regarding authoritative power and its lack of advocacy, notions of participatory planning and deliberation as being a key aspect of effective urban regeneration began to take root. Effectively these positions argued against absolutist, expert driven, knowledge and highlighted, along with other discourses (Ife and Tesoriero 2006; Kenny 2006; Kenny and Clarke 2010; King and Cruickshank 2012), that local knowledge may be just as significant (Levy 2009). These tenets were formalised in Davidoff's (1965) deliberative planning guide and Arnstein's (1969) much referenced ladder of participation which respectively argued for a social and cultural turn in planning and provided a metric for illustrating the various levels of engagement, with the lowest rung of the ladder being the pacification of the population and the highest being complete citizen control of the planning process. These concepts have grown to become industry standards in their own right, producing standardised tables exploring levels of engagement (IAP2 2007) and a significant body of work on the benefits of communicative deliberation, or the power of plural negotiations and mutual education to resolve problematic planning issues (Healey 1992).

This lean towards communication and education has not just come from theoretical positions, with proof of the strength of stakeholder engagement practices also coming from the field. Successes have been noted in Seattle (Sirianni 2008), Salt Lake City (de Souza Briggs 2008), Boston, Chicago (Ramasubramanian 2010) and Portland (Irazabal 2005) amongst others, with the result that the Grattan Institute (a peak body for social and economic research in Australia) noted that successful projects in politically, socially or culturally problematic redevelopments, internationally, can be directly attributed to early and prolonged engagement (Kelly 2010: 4). The benefits of engagement are not only achieving consensus, and therefore alleviating potential conflict from community groups, reducing political infighting and working around 'wicked' problems, but also using interest groups and small-scale democratic process to imaginatively resolve complex and intractable problems.

This is where the problem solving aspects of engagement emerge, from the 'swarm'-like activity as described by Roggema (Roggema and van der Dobbelsteen 2008) and the way in which democracy can be used as a problem solving mechanism – utilising the discursive aspect of interdisciplinary and multi-perspective negotiations to imaginatively resolve disputes and drive progressive and mutually beneficial planning schemes (de Souza Briggs 2008). Hartz-Karp's involvement in the city of Perth's strategic planning exercises provides a good local example of this where she, in consultation with the Western Australian planning minister, began a large-scale consultation process involving over a thousand participants selected from politics, industry and the general community to resolve the city's planning priorities and strategic directions over the next thirty years which produced Perth: The Network City (Hartz-Karp 2005; Hartz-Karp and Briand 2009).

One of the primary mechanisms in this process was to bring large amounts of data (both historical and future projections) together in an easy to analyse fashion, allowing users to see the effects of various scenarios and the potential futures available to them. As with other engagements, GIS was utilised to satisfy this function.

3. GIS AND GEO-TOOLS AS ENGAGEMENT AND COMMUNICATIVE TOOL

The success of GIS as a tool for stakeholder engagement can be seen in its inclusion in planning engagement praxis handbooks, land use suitability and developments in software design (Foth et al. 2009; Gordon et al. 2011; Hanzl 2007; McCall and Dunn 2012; Nedovic-Bubic 2000; Ramasubramanian 2010; Sui 2008; Walters 2007; Wates 2000). Malczewski's (2004) review of GIS being used for land suitability, through dated, provides a sound overview of not only the possible technologies and algorithms to be used, but also their ability and power to engage with stakeholders. Covering tools that range from the simple to the advanced, he illustrated that it is not necessarily the sophistication of the tool, but its ability to be easily used by stakeholders that is of most importance. This position was earlier put forward by Klosterman (1999) whose philosophy of simplicity, elegance and intuitive design led to the creation of the "What if?" system. This land use suitability tool incorporated small sets of context relevant (stakeholder and locale) parameters and, through user community engagement, allowed those ultimately affected by changes in land use to observe the various scenarios available to them.

This influential approach began the proliferation of GIS systems being used across the breadth of land development, resulting in these systems being taken up en masse as stakeholder engagement tools and practices throughout America and Europe (Ramasubramanian 2010). However, the take-up of significant levels of stakeholder engagement, and GIS as a way to achieve it, in Australia has been low (Eversole 2012; Kelly 2010; King and Cruickshank 2012; Mowbray 2005; Ramasubramanian 2010), notwithstanding some partial successes (Ghani 2011; Pettit et al. 2004) and a reasonably good supply of map based government services.

As a way forward in this area, AURIN (Australian Urban Research Infrastructure Network), a federally funded project to provide a data and e-tool hub for researchers, has proposed a set of urban research tools. One of the tools they are currently building is a nationwide version of Klosterman's "What if?" tool (Nino-Ruiz et al. 2011) where users will be able to employ community engagement philosophies for land use suitability analysis while simultaneously having access to the best available data. This group are also in the process of designing walkability, health and utility analysis tools, as well as implementing, on a national scale, the ENVISION tool for redevelopment precinct identification.

4. ENVISION

The mandate of ENVISION was to provide a platform to unite, analyse and view a wide range of data relevant to urban redevelopment which would ultimately be used as a stakeholder engagement tool to extract the tacit information held by industry experts, government and community interest groups. Though ultimately failing in its plan to generate a federated and self-updating data backbone (due to non-contiguous government data, limited services for automatically updating land data, lack of consistent protocols for compiling government data, and legal issues concerning data ownership and privacy), the system did manage to incorporate geographical data, valuations data, demographic data, information from hard and soft infrastructure, distance data and other information pertinent to the various stakeholder arenas into a usable an intuitive interface/database.

Two test cases were proposed, the City of Manningham in Melbourne, Victoria and the City of Canning in Perth, Western Australia, both of which provided access to their land data and were instrumental in the development of the system. Further data was provided by the Departments of Planning in both states. Funding came largely from the CRCSI (Cooperative Research Centre for Spatial Information), as well as annual funding inputs from state and local governments.

The package currently consists of four tools, two of which relate to stakeholder engagement and two which relate to housing capacity and density calculations. Only the first two will be included in this paper.

4.1. THE PLANNING/MCE TOOL

The first tool is the planning/MCE (multi-criteria evaluation) tool. It was designed to encourage stakeholder interaction and discussion on land use; in particular, to determine what areas to redevelop and what areas to leave out of redevelopment plans. In workshops, users selected which variables they deemed to be significant for redevelopment (such as proximity to services, transport, ages of dwellings, market effects or demographics) and then to weight these variables (1 being mildly significant and 20 being very significant). Weights were proportioned to each variable and scores were then calculated on a cadastral basis. This produced a map of municipal properties achieving high and low scores based on the query; illustrating, based on the variables and weights chosen, the areas of redevelopment focus.

The image below comes from an engagement meeting with the City of Manningham where individuals from statutory planning, strategic planning, valuations, transport and sustainability departments were present. At one point the discussion turned to aged care and where to house the elderly. The interface shows that areas with high aged demographics were selected along with proximity to shops, public transport and parks, all of which, through discussions regarding mobility, aesthetics and probability of successful engagement with the elderly, were weighted. The resultant map indicates the areas that were calculated to best adhere to the entered specifications (with paler areas being a positive outcome and darker being a negative).

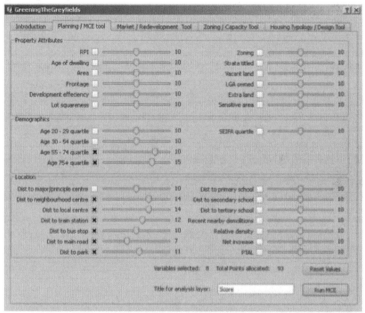

Figure 1. MCE tool with focus on aged demographics and proximity to services

Other queries involved analysing the best areas for large-scale redevelopment, proximity to schools (for little or no redevelopment) and student housing.

The effect of this tool was to allow individuals from a variety of areas within the local government to begin interdepartmental negotiations and come to consensus, with regard to proposed land use, over a series of workshops with the software. Ultimately this tool will aid in the zoning of areas where, by illustrating the effect of multiple criteria, it can highlight those that adhere to all, or most of the criteria placed upon them. Also, by showing the effect of multiple criteria, individuals can see, live, the effect of their negotiations which, due to its ability to geographically represent arguments, aids in discussions and compromise.

Figure 2. MCE map of redevelopment focus

4.2. THE MARKET REDEVELOPMENT TOOL

The second tool is less strategic and focuses more on individual cadastral redevelopment potential, rather than an potential rezoning schemes. The aim was to actually identify the dwellings that are likely to be demolished and/or be of interest to building developers. A series of variables were presented to users, each with a specific cut-off. The tool isolated the cadastres that satisfy the criteria supplied by users, with the aim of drilling down into the data and selectiv ely removing more and more properties until redevelopment precincts are identified. In the example below (also from City of Manningham workshops) the factors that were selected were a high RPI (Redevelopment Potential Index – an index of capital improved value to land value, which effectively shows the amount of value that is in the land; a value of 1 indicates that the dwelling has no value and if sold has a high probability of being demolished and redeveloped), age of dwelling over 45 (the municipal mean age of demolition) and area where there has been a significant amount of demolitions and net increase in dwellings (or areas that are currently being redeveloped).

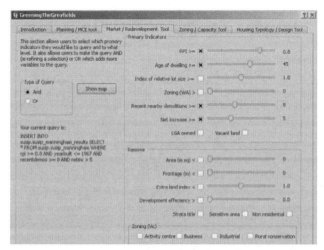

Figure 3. Redevelopment tool with high Redevelopment Potential Index, age over 45 and in areas of moderate redevelopment activity selected

The resultant map from these queries shows pockets of high redevelopment potential (the dark areas) in the north-west, south-west, central and south-eastern corner of the municipality, most of which lie in highly redevelopable areas (as identified from the earlier query).

Figure 4. Map of highly redevelopable dwellings

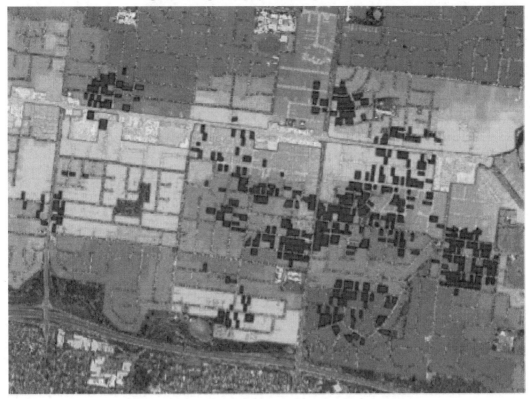

Figure 5. Close-up of potential redevelopment precincts

When zoomed in, the map clearly identifies a potential redevelopment precinct that coinsides with the strategic redevelopment area, particularly the precinct in the most north-east area of the map. Note also that the map identifies cadastres that could potentially be consolidated, thus leading to the more large-scale development associated with aged care (or other precinct style) constructions.

The result from this single workshop was the identification of this precinct by the multi-disciplinary panel. This further led to debate over extending the current zoning practices to include these precincts and the potential for including them in future, higher density, redevelopment zones.

5. RESULTS OF PARTICIPANT ENGAGEMENT

5.1. COLLABORATIVE SOFTWARE DESIGN

The initial specifications for the tool were very loose, basically that it be a GIS platform for viewing and drilling down into large amounts of land data for identifying redevelopment precincts. The tool began as a front end to a geo-database. However, due to early stakeholder engagement, it quickly became a number of specific tools, each with their own function. Interaction with the Western Australian test site (the City of Canning) revealed the need for a multi-criteria evaluation tool, with the variables for the system coming directly from the strategic developers in the municipality. Collaboration between the Western Australian (Curtin) and Victorian (Swinburne) universities involved in the project led to a negotiated set of data common to both states and based on a composite of available data sets. A similar process led to the market based tool, with collaboration, stakeholder engagement and availability of data sets resulting in a tool that could be used in both states. The rezoning tool (not used in this paper) was developed purely for a housing capacity analysis for the City of Canning, while the amalgamation tool (also not used in this paper) grew out of the need to populate precincts and was based on collaboration with the architecture department at Monash University. Further engagement led to stakeholders asking for additions to the package, such as photo imagery, Google maps, slope of land, rental properties and reporting functions, all of which have been, or are scheduled to be, implemented. As such, the development of the set of tools was largely informed by collaboration between both states the project was running in and, more importantly, directly engaging with stakeholders as to their requirements.

Currently the software is about to be adapted and brought into the AURIN portal. As a by-product of presenting this software to AURIN, similar projects at the University of Melbourne have shown interest in combining their models with ENVISION to produce a suite of generic products that can be utilised across the urban development spectrum. As this process is already aiding in the development of mutually beneficial protocols and advances to both sets of modelling tools, it is anticipated that these discussions will lead to further developments in the software.

This form of cyclical development based on fast feedback loops concurs with AGILE development methodologies (Beck et al. 2001), however, when used in tandem with broad spectrum and wide-ranging stakeholder engagement it produces software that is collaboratively designed not just by one stakeholder, but by the range of stakeholders involved in the planning environment.

5.2. COMMUNICATIVE ENGAGEMENT

Newton, Newman et al. (2012) have identified ten key stakeholder engagement arenas in the urban regeneration area. The three that have thus far been engaged are internal state government, internal local government and the relationships between state and local government (the discussions between property developers and community members are forthcoming).

5.2.1. STATE GOVERNMENT ENGAGEMENT

The discussions and workshops with state government involved directors and planners from strategic planning, statutory planning, policy development, activity centre development, transport planning, urban regeneration and urban growth development. Initial discussion highlighted the need for ENVISION to produce reports and capture data on the effects of redevelopment, however, these discussions also grew into exploring the power of the tool to be used across many municipalities. Later workshops with different sets of government stakeholders produced similar results, with the general consensus being reached that tools such as ENVISION could not only be used for policy change (by illustrating the effect of business-as-usual practices versus more advanced designs) but more broadly as a form of meta-governance to provide the tools needed by local governments to achieve state strategies. In short, the political effect of the workshops was to highlight, amongst various sectors of state governance, the need for tools and policy-led solutions towards achieving the goals set out in the Greening the Greyfields project.

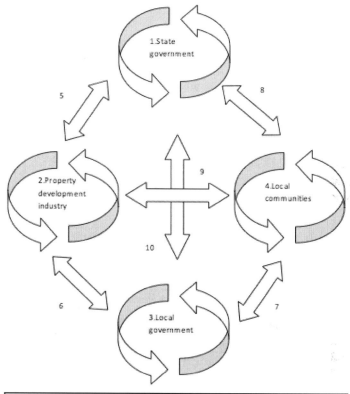

1. State governments (planning and related departments and appeals tribunals etc)
2. Property development industry (for profit; not for profit)
3. Local government (council officers; elected councillors etc)
4. Local community (municipal: specific neighbourhoods etc)
5. Envisioning future redevelopment – major projects; planning appeals
6. Envisioning future redevelopment – individual projects (pre-planning permits)
7. Community engagement; long-term development strategy
8. Community engagement; long-term development planning
9. Brokering precinct regeneration
10. Envisioning and agreeing future development strategies.

Figure 6. Stakeholder Engagement Arenas for Urban Development

Some of the policy changes that were suggested included a move away from concentration on individual lots to looking at how best to consolidate lots. A lot amalgamation bonus or tax was suggested, as were policies based on land use education for key stakeholders. It was also suggested that municipalities be responsible for reporting on their strategic targets, saying how they were going to implement the targets set by state. The effect would be to encourage the use of more strategic tools which would ultimately better educate both the planners and the community members as to the potential of different redevelopment options.

A second response was how the workshop attendees supported the tool and actively looked for ways that it could be used. This debate stimulated ideas amongst, getting them to think of novel ways to resolve key planning issues. Most saw the tool as a way to represent large amounts of data from a wide range of data sets in a simple and usable form, indicating that with tools such as ENVISION, different groups within (and without) government could see, real time, the data pertinent to their own position as well as the pertinent data of other groups, making cross-disciplinary communication easier. It was also noted by many attendees that the tool's ability to look at precinct and municipalities (as opposed to just single cadastres or projects) made strategic planning with the tool a distinct possibility. One pertinent suggestion was that the tool be used to tackle the Norlane project, a social housing redevelopment project which has been on the cards as a state redevelopment project for at least ten years. It was put forward that ENVISION could be used as a way to present the complete range of options to Office of Housing managers (who own 50% of the stock in these suburbs), developers and community members. Another was a way to implement the current state changes in zoning, with the tool being capable of showing where potential redevelopment should be placed.

The final result at the state level, though largely unmeasurable, was the positive effect that the workshops had on the mentalities of those attending them. Though many attendees were initially quite cynical about "yet another urban redevelopment tool", as discussion started to take place, cross-disciplinary ideas came forward and novel ideas started to flow, the inertia of large organisational and silo based thinking began to dissolve. People started to become quite passionate about their ability to make positive change and, being free to image potential futures (outside of current limitations), began to come up with novel ideas and solutions to some of the key issues with planning. In effect, the free-thinking and discursive aspect of the workshops allowed individuals, using the tool as a focus, to become quite creative in how to solve problems that were previously intractable.

In sum, the state workshops illustrated that:

1. There exists within state government the ability to creatively resolve complex planning issues, but this is hampered by institutional inertia, political difficulty and the lack of space for creative freedom coming from cross-disciplinary discussion within and across key departments/offices.
2. Multi-stakeholder collaboration allows for diverse positions to pool their collective resources and imaginatively resolve complex and wicked planning issues.
3. Policy shift is the key mechanism for changing the future, but these policies rest on the ability of the respective departments to envision a positive future and to engage with how best to achieve these ends (not simply focusing on the restrictions and practicalities of current systems)
4. There is currently little discussion between state and local government which needs to occur for effective governance in the area of strategic planning to take effect.
5. Meta-governance and meta-governance tools are required by municipalities. These should come in the form of policy change, but also in the form of strategic tools such as ENVISION, which allow locales the power to decide their future, but with the data consolidation and communicative tools that ENVISION is emblematic of.

5.2.2. LOCAL GOVERNMENT ENGAGEMENT

Engagement workshops with local government produced the same levels of interdisciplinary discussion, allowing those involved in government functions, including transport, sustainability, strategic and statutory planning, valuations, water, electricity and other services, to come together and discuss possibilities for the future. As with the state workshops these events brought an element of passionate creativity to the group as they could see, real time, the effects of their choices. Once the basic framework was demonstrated, groups started to identify potential redevelopment precincts, as well as attempting to use the tool to focus on where developments should not occur, where market pressures were strongest and potential changes to their current zones.

There were a number of positive outcomes and lessons learnt from these workshops. The first was the power of community members to influence strategic decisions, with numerous stories emerging illustrating how zoning changes and large, or even small, development projects had been hampered by concerned locals. This was all said with no mention of actual or historical community engagement, which further illustrated the lack of community consultation, aside from presentation of pre-prepared plans to community members. The reciprocal of this (not being informed, involved in or adequately assisted by state government) was also a pertinent issue which highlighted the necessity of consultation between state and local governments, and which ENVISION workshop leaders and project members are now ideally placed to do.

The issue of policy change, particularly for land accumulation, aged housing, housing affordability and houses for young families, was raised a number of times. Local community members saw these strategies as not necessarily coming from the state, but could be implemented locally through community education and locally trying to lead new projects aimed at medium level redevelopment. It was felt that with a combination of localised policies, education and a significant amount of lead time, the local government could achieve a moderate and acceptable level of redevelopment that would be accepted by community members.

Parking, transport and accessibility were also continually raised, with the traffic engineers and those in the front line of development application processing noting that without adequate provision for transport, congestion and garage space, plans would not go ahead. Planners also argued that storage space (particularly for those moving to medium density developments) was, after parking, the most important issue for locals. This discussion led to the issue of the development typologies that were not only acceptable, but also appropriate for the locale, which also included presentation of the typology work coming from Monash University.

Finally, It became apparent that the current design and development overlays (the areas zoned for medium density development) were already exhausted, having been mostly redeveloped, but to no significant level. This raised the issue of how to move forward, in terms of redeveloping the locale while not raising the ire of locals. While rezoning was being workshopped it was also noted that the tool could also be used to allay the fears of locals who were worried about mass development by illustrating the lack of market pressures in contended areas, and therefore the low probability of large-scale redevelopment in their 'back yards'.

The result of local government engagement workshops was that they:

1. Identified the need for engagement with both community members and state government;
2. Highlighted the need for policy change, with community education done locally and meta-governance from the state to provide the mechanisms for doing so;
3. Showed the need for tools to aid in the amassing of data to assist them with the development of future zones;
4. Illustrated the plural approach that is required when attempting to resolve complex issues;
5. Identified possibly redevelopment precincts, or at least now understand that they have the potential to do so in the future.

In sum, ENVISION provided the focal point for numerous individuals from a variety of sectors to engage in a common issue and to observe the result of their comments real time. It showed the potential of redevelopment, both within current constraints and when they are loosened. Though only having engaged in three of the ten stakeholder arenas thus far, the tool has already highlighted the power of GIS systems to not only educate but to act as a tool for negotiation between diverse positions.

6. THE NEXT STEPS

The next step for ENVISION will be its inclusion in the AURIN data hub and set of urban research e-tools where it will be made accessible to the wider urban development community, using the two existing data sets (the municipalities of Canning and Manningham) as test cases. Given the development time for this exercise (one year) and the existence of walkability, health, sustainability and other urban metrics already within the AURIN portal, it is assumed that these will inform the next iteration of the software, where it will begin to incorporate a diverse array of hard and soft infrastructure feedback loops which will further enrich the ability of the tool to connect with different voices in the urban redevelopment arena.

The funding round for the next phase of the Greening the Greyfields project has also begun where we will be taking the tool and adding 3D precinct visualisation, additional feedback mechanisms and redevelopment typologies to it, making it into a complete stakeholder, modelling and analysis tool for urban regeneration and community engagement.

7. CONCLUSION

In the face of continued urban expansion, the increase in individuals living in urban environments and, in the Australian context, the unsustainable way in which this is being strategically managed, there is a growing call for technological and planning reform, based on multi-level stakeholder engagement and data integration, which Greening the Greyfields and ENVISION are primarily concerned with. Though only part way through the project/software development process, there has already been tremendous success in terms of acquiring insider knowledge, obtaining government endorsement and negotiating further investment. This success has come from the project's focus on collaborative and communicative engagement, through obtaining the wide-ranging data and providing the interface for individuals to manipulate it according to their opinions on future planning schemes. Effectively the tool, and the way that it allows experts to educate each other and potentially engage with the local community, has shown the powers of geography and geographical information tools. In providing a focus for discussion, as well as displaying results of discourse specific queries instantaneously, it has proved to be an extremely effective mechanism for broad-spectrum multi-stakeholder engagement.

REFERENCES

Arnstein, S.R. 1969. A Ladder of Citizen Participation. Journal of the American Institute of Planners:35(4): 216-224.

Aulich, C. 2009. From Citizen Participation to Participatory Governance in Australian Local Government Commonwealth Journal of Local Governance:(2): 44-60.

Manifesto for Agile Software Development website 2001. Available from: http://agilemanifesto.org/ [Accessed Sept. 2012].

Brody, S.D., D.R. Godschalk, and R.J. Burby 2003. Mandating Citizen Participation in Plan Making: Six Strategic Planning Choices. Journal of the American Planning Association:69(3): 245-264.

Carson, L. 2011. Dilemmas, Disasters and Deliberative Democracy. Griffith Review:32: 38-46.

COAG Reform Council 2012. Review of Capital City Strategic Planning Systems. Sydney: COAG Reform Council.

Creighton, J. 2005. The Public Participation Handbook: Making Better Decisions Through Citizen Involvement. San Francisco: John Wiley & Sons.

Davidoff, P. 1965. Advocacy and Pluralism in Planning. Journal of the American Institute of Planners:31(4): 331-338.

de Souza Briggs, X. 2008. Democracy as Problem Solving: Civic Capacity in Communities Across the Globe. London: MIT Press.

Department of Planning and Community Development 2012. Metropolitan Planning Strategy: A Vision for Victoria (Information on Strategic Principles). Melbourne: Department of Planning and

Community Development.

Eversole, R. 2012. Remaking Participation: Challenges for Community Development Practice. Community Development Journal 47(1): 29-41.

Foth, M., et al. 2009. The Second Life of Urban Planning? Using NeoGeography Tools for Community Engagement. Journal of Location Based Services 3(2): 97-117.

Fritze, J.,. Williamson, L., Wiseman J. 2009. Community Engagement and Climate Change: Benefits, Challenges and Strategies. Melbourne: Department of Planning and Community Development.

Gallent, N., Robinson S. 2012. Neighbourhood Planning: Communities Networks and Governance. Bristol: Polity.

Geddes, M. 2006. Neo-liberalism and Local Governance: Cross-National Perspectives and Speculations. Policy Studies:36(3:4): 359-377.

Ghani, M.Z.A. 2011. Virtual Werribee. 19th International Conference of Modelling and Simulation. Perth.

Gordon, E., Schirra, S. , Hollander J.2011. Immersive Planning: A Conceptual Model for Designing Public Participation with New Technologies. Environment and Planning B: Planning and Design:38: 505-519.

Hanzl, M. 2007. Information Technology as a Tool for Public Participation in Urban Planning: A Review of Experiments and Potentials. Design Studies 28: 289-307.

Hartz-Karp, J. 2005. A Case Study in Deliberative Democracy: Dialogue with the City. Journal of Public Deliberation:1(1).

Hartz-Karp, J., Briand M.K. 2009. Practitioner Paper: Institutionalising Deliberative Democracy. Journal of Public Affairs:9: 125-141.

Healey, P. 1992. Planning Through Debate: The Communicative Turn in Planning Theory. Town Planning review:63(2): 143-162.

Herriman, J. 2011. Local Government and Community Engagement in Australia. Working Paper No 5. Australian Centre of Excellence for Local Government: University of Technology Sydney.

IAP2 Spectrum of Public Participation website 2007. Available from: http://www.iap2.org/associations/4748/files/spectrum.pdf [Accessed August. 2012].

Ife, J., Tesoriero F. 2006. Community Development:Community Based Alternatives in an Age of Globalisation. Sydney: Pearson Education Australia.

Innes, J.E., Booher D.E. 2011. Planning with Complexity: An Introduction to Collaborative Rationality for Public Policy. New York: Routledge.

Irazabal, C. 2005. City Making and Urban Governance in the Americas: Curtiba and Portland. Aldershot: Ashgate.

Jarvis, D., Berkeley, N. Broughton K. 2012. Evidencing the Impact of Community Engagement in Neighbourhood Regeneration: The Case of Canley, Coventry. Community Development Journal:47(2): 232-247.

Kelly, J.F. 2010. Cities: Who Decides? Melbourne: The Grattan Institute.

Kenny, S. 2006. Developing Communities for the Future. Melbourne: Thompson.

Kenny, S., Clarke M. 2010. Challenging Capacity Building: Comparative Perspectives New York: Palgrave Macmillan.

King, C., Cruickshank M. 2012. Building Capacity to Engage: Community Engagement or Government Engagement? Community Development Journal:47(1): 5-28.

Klosterman, R.E. 1999. The What if? Collaborative Support System. Environment and Planning B: Planning and Design:26: 393-408.

Lange, E. 2011. 99 Volumes Later: We Can Visualise. Now What? Landscape and Urban Planning:100: 403-406.

Lawson, L., Kearns A. 2010. Community Engagement in Regeneration: Are We Getting the Point? Journal of Housing and the Built Environment:25: 19-36.

Le Dantec, C.A. 2012. Participation and Publics: Supporting Community Engagement CHI 2012. Austin, Texas.

Levy, J.M. 2009. Contemporary Urban Planning. Upper Saddle River, NJ: Pearson Prentice Hall.

Malczewski, J. 2004. GIS-Based Land-Use Suitability Analysis: A Critical Overview. Progress in Planning:62: 3-65.

McCall, M., Dunn C. 2012. Geo-Information Tools for Participatory Spatial Planning: Fulfilling the Criteria for ‚Good' Governance? Geoforum:4: 81-94.

Mowbray, M. 2005. Community Capacity Building of State Opportunism? Community Development Journal:40(3): 255-264.

Murayama, A. 2008. Toward the Development of Plan-Making Methodology for Urban Regeneration. In: H. Massahide, and K. Hideki, eds. Innovations in Collaborative Urban Regeneration. New York: Springer, 15-29.

Nedovic-Bubic, Z. 2000. Geographic Infomation Science Implications for Urban and Regional Planning. URISA Journal:12(2): 81-93.

Newton, P. 2010. Beyond Greenfield and Brownfield: The Challenge of Regenerating Australia's Greyfield Suburbs. Built Environment:36(1): 81-103.

Newton, P., et al. 2011. Towards a New Development Model for Housing Regeneration in Greyfield Residential Precincts. Melbourne: Australian Housing and Urban Research Institute (AHURI).

Newton, P., et al. 2012. Greening the Greyfields: Unlocking the Redevelopment Potential of the Middle Suburbs in Australian Cities. International Conference on Urban Planning and Regional Development. Venice.

Nino-Ruiz, M., et al. 2011. AURIN What If?: Decision Support for Projections of Land Use Allocations. 5th eResearch Australasia Conference. Melbourne.

NSW Government 2012. A New Planning System for NSW - Green Paper. Sydney: NSW Government.

OECD 2012. Compact City Policies: A Comparative Assessment. OECD Green Growth Studies: OECD Publishing.

Pettit, C., A. Nelson, and W. Cartwright 2004. Using On-Line Geographical Visualisation Tools to Improve Land Use Decision-Making with a Bottom-Up Community Participation Approach. In: J.P. van Leeuwen, and H.J.P. Timmermans, eds. Recent Advances in Design and Decision Support Systems in Architecture and Urban Planning. London: Kluwer, 53-68.

Ramasubramanian, L. 2010. Geographic Information Science and Public Participation. Dordrecht: Springer.

Reddel, T., Woolcock G. 2004. From Consultation to Participatory Governance? A Critical Review of Citizen Engagement Strategies in Queensland. Australian Journal of Public Administration:63(3): 75-87.

Roberts, N. 2000. Wicked Problems and Network Approaches to Resolution. International Public Management Review:1(1): 1-19.

Roggema, R., Van der Dobbelsteen A.2008. Swarm Planning: Development of a New Planning Paradigm. World Sustainability Conference. Melbourne.

Russell, J.S. 2011. The Agile City: Building Well-being and Wealth in an Era of Climate Change. Washington: Island Press.

Sanoff, H. 2000. Community Participation Methods in Design and Planning. New York: John Wiley & Sons.

Sirianni, C. 2008. Investing in Democracy: Engaging Citizens in Collaborative Governance. Washington: Brookings Institution Press.

Smyth, P.,. Reddel, T ,. Jones A.E.G. 2005. Community and Local Governance in Australia. Sydney: UNSW Press.

Sorensen, E.,. Torfing, J. eds. 2007. Theories of Democratic Network Governance: Palgrave Macmillan.

Sui, D.Z. 2008. The Wikification of GIS and Its Consequences: Or Angelina Jolie's New Tattoo and the Future of GIS Computers, Environment and Urban Systems:32(1): 1-5.

Walters, D. 2007. Designing Community: Charrettes, Masterplans and Form-Based Codes. Oxford: Elsevier.

Wates, N. 2000. The Community Planning Handbook. London: Earthscan.